Titles include:

Scott Black
OF ESSAYS AND READING IN EARLY MODERN BRITAIN

Claire Brock
THE FEMINIZATION OF FAME, 1750–1830

Brycchan Carey
BRITISH ABOLITIONISM AND THE RHETORIC OF SENSIBILITY
Writing, Sentiment, and Slavery, 1760–1807

E. J. Clery
THE FEMINIZATION DEBATE IN 18TH-CENTURY ENGLAND
Literature, Commerce and Luxury

Adriana Craciun
BRITISH WOMEN WRITERS AND THE FRENCH REVOLUTION
Citizens of the World

Ildiko Csengei
SYMPATHY, SENSIBILITY AND THE LITERATURE OF FEELING IN THE
EIGHTEENTH CENTURY

Peter de Bolla, Nigel Leask and David Simpson (*editors*)
LAND, NATION AND CULTURE, 1740–1840
Thinking the Republic of Taste

Elizabeth Eger
BLUESTOCKINGS
Women of Reason from Enlightenment to Romanticism

Ina Ferris and Paul Keen (*editors*)
BOOKISH HISTORIES
Books, Literature, and Commercial Modernity, 1700–1900

John Gardner
POETRY AND POPULAR PROTEST
Peterloo, Cato Street and the Queen Caroline Controversy

George C. Grinnell
THE AGE OF HYPOCHONDRIA
Interpreting Romantic Health and Illness

Ian Haywood
BLOODY ROMANTICISM
Spectacular Violence and the Politics of Representation, 1776–1832

Anthony S. Jarrells
BRITAIN'S BLOODLESS REVOLUTIONS
1688 and the Romantic Reform of Literature

Jacqueline M. Labbe
WRITING ROMANTICISM
Charlotte Smith and William Wordsworth, 1784–1807

Michelle Levy
FAMILY AUTHORSHIP AND ROMANTIC PRINT CULTURE

April London
LITERARY HISTORY WRITING, 1770–1820

Robert Miles
ROMANTIC MISFITS

Tom Mole
BYRON'S ROMANTIC CELEBRITY
Industrial Culture and the Hermeneutic of Intimacy

Catherine Packham
EIGHTEENTH-CENTURY VITALISM
Bodies, Culture, Politics

Nicola Parsons
READING GOSSIP IN EARLY EIGHTEENTH-CENTURY ENGLAND

Jessica Richard
THE ROMANCE OF GAMBLING IN THE EIGHTEENTH-CENTURY BRITISH NOVEL

Andrew Rudd
SYMPATHY AND INDIA IN BRITISH LITERATURE, 1770–1830

Erik Simpson
LITERARY MINSTRELSY, 1770–1830
Minstrels and Improvisers in British, Irish and American Literature

Anne H. Stevens
BRITISH HISTORICAL FICTION BEFORE SCOTT

David Stewart
ROMANTIC MAGAZINES AND METROPOLITAN LITERARY CULTURE

Mary Waters
BRITISH WOMEN WRITERS AND THE PROFESSION OF LITERARY CRITICISM,
1789–1832

P. Westover
NECROMANTICISM
Travelling to Meet the Dead, 1750–1860

Esther Wohlgemut
ROMANTIC COSMOPOLITANISM

David Worrall
THE POLITICS OF ROMANTIC THEATRICALITY, 1787–1832
The Road to the Stage

**Palgrave Studies in the Enlightenment, Romanticism and Cultures of Print
Series Standing Order ISBN 978–1–4039–3408–6 hardback 978–1–4039–3409–3
paperback** (*outside North America only*)

You can receive future titles in this series as they are published by placing a standing
order. Please contact your bookseller or, in case of difficulty, write to us at the address
below with your name and address, the title of the series and the ISBN quoted
above.

Customer Services Department, Macmillan Distribution Ltd, Houndmills, Basingstoke,
Hampshire RG21 6XS, England

Eighteenth-Century Vitalism

Bodies, Culture, Politics

Catherine Packham

First published 2012 by
PALGRAVE MACMILLAN

Palgrave Macmillan in the UK is an imprint of Macmillan Publishers Limited, registered in England, company number 785998, of Houndmills, Basingstoke, Hampshire RG21 6XS.

Palgrave Macmillan in the US is a division of St Martin's Press LLC, 175 Fifth Avenue, New York, NY 10010.

Palgrave Macmillan is the global academic imprint of the above companies and has companies and representatives throughout the world.

Palgrave® and Macmillan® are registered trademarks in the United States, the United Kingdom, Europe and other countries.

ISBN 978–0–230–27618–5

This book is printed on paper suitable for recycling and made from fully managed and sustained forest sources. Logging, pulping and manufacturing processes are expected to conform to the environmental regulations of the country of origin.

A catalogue record for this book is available from the British Library.

A catalog record for this book is available from the Library of Congress.

Contents

List of Illustrations   vi

Acknowledgements   vii

Introduction   1

Part I   Writing the Body in the Scottish Enlightenment

1   Forms of Enlightenment: Embodied Beings in
Eighteenth-Century Scotland   25

2   Generating Sympathy: Sensibility, Animation and Vitality
in Adam Smith and Mary Wollstonecraft   52

3   Labouring Bodies in Political Economy: Vitalist Physiology
and the Body Politic   83

Part II   Enlightenment in the 1790s: The Scottish Legacy

4   Enlightenment Legacies and Cultural Radicalism: Physiology
and Politics in the 1790s   111

Part III   Vitalism, Animation, Culture

5   Animated Nature: Erasmus Darwin and the Poetry and Politics
of Vital Matter, 1789–1803   147

6   Animation and Vitality in Women's Writing of the 1790s   175

Conclusion: Eighteenth-Century Vitalism, Romanticism,
Literature and the Disciplines   207

Notes   217

Bibliography   237

Index   247

List of Illustrations

1. James Gillray, *An Excrescence; – a Fungus; – Alias – a Toadstool upon a Dung-Hill.* © The Trustees of the British Museum. Reproduced by permission of the British Museum, PRN. PPA139467   123

2. James Gillray, *The Giant-Factotum Amusing Himself.* © The Trustees of the British Museum. Reproduced by permission of the British Museum, PRN.PPA142915   124

3. James Gillray, *New Morality; – or – the promis'd instalment of the high-priest of the Theophilanthropes, with the homage of Leviathan and his suite, Anti-Jacobin Review and Magazine; or Monthly Political and Literary Censor,* no. 1 (1 August 1798), 1. © The British Library Board. Reproduced by permission of the British Library, cat. no.261.i.I   163

4. James Gillray, *The Apples and the Horse-Turds; – or – Buonaparte among the golden pippins.* © The Trustees of the British Museum. Reproduced by permission of the British Museum, PRN. PPA144262   164

5. Frontispiece to Mary Ann Radcliffe, *The Female Advocate,* 1799. © The British Library Board. Reproduced by permission of the British Library, cat. No C. 123.f.17   180

Acknowledgements

Portions of this book have appeared in earlier form in the following publications. The last section of Chapter 1 appeared as 'Disability and Sympathetic Sociability in Enlightenment Scotland: The Case of Thomas Blacklock', *British Journal for Eighteenth-Century Studies*, 30:3 (2007), 423–38. Reproduced by permission of John Wiley and Sons. Material from Chapter 3 appeared as 'The Physiology of Political Economy: Vitalism and Adam Smith's *Wealth of Nations*', *Journal of the History of Ideas*, 63:3 (2002), 465–81. Reproduced by permission of the University of Pennsylvania Press. Chapter 5 draws on material previously published in 'The Science and Poetry of Animation: Personification, Analogy, and Erasmus Darwin's *Loves of the Plants*', *Romanticism*, 10:2 (2004), 191–208, and appears here by permission of the editor. Some material from 'Feigning Fictions: Imagination, Hypothesis, and Philosophy in the Scottish Enlightenment', *Eighteenth Century: Theory and Interpretation*, 44:2 (2007), 149–71, appears in the Introduction to this book. Reproduced by permission of Texas Tech University Press.

The origins of this book lie in a doctoral thesis on Adam Smith supervised by Peter de Bolla, to whose expert guidance and encouragement I am greatly indebted. At the School of English, Leeds University, I was lucky to participate for a number of years in a stimulating community of scholars and students of the eighteenth-century and Romantic periods, including David Fairer, Robert Jones, Nick Seager and Vivien Jones. Also at Leeds, John Whale was a supportive and generous research mentor whose energies and insight were much valued as this book took shape. I'm grateful to the Arts and Humanities Research Council and to the School of English, University of Sussex, for funding a precious year's research leave which enabled me to complete the first draft of the book, and to John Barrell, and my then Head of School, Andrew Hadfield, for supporting that application. I would like to thank also the two anonymous readers for Palgrave Macmillan, whose comments and criticisms have made this a much better book than it otherwise might have been. Any errors, omissions or oversights of course remain my own. At Palgrave Macmillan, Paula Kennedy and Ben Doyle have overseen the book's production with admirable efficiency. Finally, I am grateful to the series editors of Palgrave Studies in the Enlightenment, Romanticism and Cultures of Print, Anne Mellor and Cliff Siskin, for their support for the project at various stages.

Thanks of a different kind go to friends and family for their numerous expressions of support and interest as the book progressed. Rebecca Beasley, Tracy Hargreaves, Kathleen Hebden, Veronica and David Packham, Emma and Ben Page, and all the Brighton gang, deserve special mentions; Dave Fairhurst deserves no thanks for setting this book's title as a charades clue during a Christmas party in Nottingham in 2010. Ed Hebden has been an unwavering source of love, sanity, patience and reassurance since long before this book was conceived; more recently, Anna and Miranda have shown me how much more there is to know about love, joy and pride. Finally, my father, with over 30 years' service to British universities, and a lifetime's consideration of questions of science, philosophy and education, and who as I write is, at age 71, walking Spain's gruelling Camino de Santiago for the second time, remains an inspirational model of critical thinking and wisdom. This book is dedicated to him.

Introduction

> And now we might add something concerning a certain most
> subtle Spirit which pervades and lies hid in all gross bodies
> Newton, General Scholium, *Principia* (2nd edition), 1713

> And what if all of animated nature
> Be but organic Harps diversly fram'd,
> That tremble into thought, as o'er them sweeps
> Plastic and vast, one intellectual breeze,
> At once the Soul of each, and God of all?
> Coleridge, *The Eolian Harp*, 1796

Vitalism, the theory that life is generated and sustained through some form of non-mechanical force or power specific to and located in living bodies, is usually associated with Romantic theories of nature. It is often connected with particular perceptions of nature: as possessing independent powers of animation and self-direction, vital energies of self-generation and the ability to take actions – whether an unconscious muscular response in an animal, or the unfurling of leaves towards the sun in a plant – which will best promote an organism's well-being. Coleridge's speculation in *The Eolian Harp* that 'all of animated nature' might 'tremble into thought', with its attribution of even further powers to nature – an embryonic consciousness – can be read as one iconic moment of such Romantic vitalism, and, indeed, Coleridge himself quotes these very lines in a letter to John Thelwall, author of a controversial essay on 'animal vitality', which reviews various contemporary theories of life and vitality. As with too many critics and historians since, Coleridge was ready to dismiss the era immediately prior to his own as an age of mechanism, but speculation about the possible, if as yet unknown powers possessed by organic matter, in fact recurred throughout the eighteenth century, and directly informed the very debates which Coleridge enthusiastically surveys.[1] Stimulated by the inability of the mechanical natural philosophy of the late

1

seventeenth century to explain the actions and functioning of living bodies, and prompted too by Newton's speculations in the 'General Scholium' to his *Principia* and in the Queries which were appended to his *Opticks*, physiologists, philosophers and experimenters grappled throughout the eighteenth century to offer alternative accounts of living nature. The supposition of a 'vital principle' or 'power', a mysterious, non-mechanical life-force whose energies animated the living world, became central to a new understanding of nature, whose self-activating powers were comprehensible neither via the laws of motion nor as directly manifesting the hand of God, but as unique to living matter. Nature, for many eighteenth-century thinkers and writers, thus emerges less as the intricate mechanism of divine clockwork, more as the repository of forces which, if unknown and unknowable in the final analysis, are nevertheless powerfully evident in innumerable observable acts of animal and vegetable growth, generation and reproduction, motion, self-preservation and self-healing. Equally distinct from animism, which explained such acts by reference to a soul, and from later, fully fledged Romantic organicism, eighteenth-century vitalism marked the transitional period between the rejection of earlier mechanical models and the formalisation of the modern sciences of life, including the discipline of biology, at the beginning of the nineteenth century.

This new 'language' of vital nature is evident across the spectrum of eighteenth-century scientific and philosophical endeavour.[2] It is also a persistent presence beyond the immediate fields of scientific enquiry. This book is a study of its manifestation, deployment and effects in a number of areas of cultural activity: in literature, poetry and fiction, but also in moral philosophy, economics and political writing. Peter Hanns Reill has described how an Enlightenment 'vitalising' of nature informed natural history, chemistry and the life sciences; its influence was felt in geology, physiology and medicine, in theories of generation, species development and ontogeny. At a time before the consolidation of divisions between what modernity regards as distinct disciplines, this sense of a newly 'vitalised' nature was felt all the more readily in closely linked areas of knowledge and enquiry. Developments in post-Newtonian matter theory in turn informed a range of experimental and theoretical sciences; sceptical philosophy played a key role in the critique of concepts central to mechanical natural philosophy, and the theological implications of the new vision of nature, which challenged Newtonian science's apparent confirmation of God's role as the universe's first and final cause, were also readily apparent. In such a context, lines such as the following, from Pope's *Essay on Man*, written between 1730 and 1732, and often understood as offering an orthodox popular account of Newtonian natural philosophy, reveal an unexpected emphasis on the sheer animate powers of material nature:

> See Matter next, with various life endu'd,
> Press to one centre still, the gen'ral Good.

> See dying vegetables life sustain,
> See life dissolving vegetate again:
> All forms that perish other forms supply,
> (By turns we catch the vital breath, and die)
> Like bubble on the Sea of matter born,
> They rise, they break, and to that sea return.
> Nothing is foreign: Parts relate to whole;
> One all-extending all-preserving Soul
> Connects each being, greatest with the least;
> Made Beast in aid of Man, and Man of Beast;
> All serv'd, all serving! nothing stands alone;
> The chain holds on, and where it ends, unknown.[3]

Form succeeds form here to articulate an overarching force of animation and generation more significant than the various parts of nature which it temporarily imbues with life. The sheer superfluity and power of organic life overpowers the reader with a sense of the same rich vitality of nature which the natural philosophy of the time was so keen to explain, but the question of the exact nature of the 'vital breath' which has animated it remains obscure. In a poem which has been described as 'strategically contradictory', Pope's depiction of the natural and infinite generation of interconnected life could be read, simply, as the celebration of a divinely created universe. But equally, the vision presented here of a natural world constantly renewing itself via thrusting, powerful energies, can be seen to anticipate the controversial poetry of Erasmus Darwin at the end of the eighteenth century, where Popean couplets expound a vision of a self-animating nature remarkably similar to that presented here.[4] Keen to expound a vision of the powers of active nature, but unclear, in the last analysis, of their ultimate nature or cause, these lines illustrate the combination of fascination and uncertainty which marked much eighteenth-century writing on nature's vitality.

Whilst vitalism can be readily identified in scientific work in this period – especially, as this book stresses, in physiology – this passage from Pope reminds us that there is across much eighteenth-century writing a recurring fascination with modes and forms of vital animation which exceeds a narrowly religious or scientific context. The relation of such writing to what can be more formally identified as vitalist theories of nature or life is the concern of the following chapters, which trace the plurality of ways, contexts and discourses in which vitalist concepts, images and languages occur in this period. This study thus understands vitalism both as a scientific theory, and in looser, more suggestive ways, in order to explore how a language of vital animation is present beyond scientific and physiological discourses, in a range of cultural, political and literary domains. References to vitality are certainly far from unusual in many kinds of writing in this period. Such references can often be understood to reflect something of the debates over the nature of life which raged in scientific and philosophical circles at this time, but they are also frequent enough to appear to have taken on (in an appropriate metaphor) a life of their own. 'Vital' spirits or sparks, vital particles or functions, and vital energies abound; they are present, for example, in Richard Steele's address, in his play *The Funeral* (1701), of the 'Vital Force, and Motion' which animates the 'Breathless lump of Clay'; in the representation, by the anonymous author of *The Ladies Dispensatory* (1740), of women as 'the Repository' of 'every original vital Particle' of life; in Cowper's description, in *The Task* (1784), of the 'vital energy' which moves the 'pure and subtile lymph' through plants; and in, additionally, more commonplace and general references to mankind's 'vital strength' or 'vital springs', and to 'vital air' and 'vital breath'.[5] The very frequency of such images should not blind us to questions about what might be being named in such phrases, which might equally suggest religious conceptions of a divine animation, or (increasingly by the end of the century) physiological notions of life-forces or living principles, uncertainly located in the air, breath, blood or body, or something somewhat hazily in between. Often such language of vitality invokes contemporary theological and philosophical conceptions only to rework them; it plays with established senses of biology and physiology, of bodies, breath and life, but pushes those senses into new meanings, as though the nature of vitality itself exceeds the ability of language to express it, or to be understood as a narrowly scientific concept. Something of this is discernible, perhaps, in the passage from Pope already quoted, and something similar can be seen at a later point in the poem, where Pope describes, in language which could be Darwin's, 'one nature' feeding a 'vital flame', an 'all-quick'ning æther' of 'life' which 'swells the genial seeds'.[6] Here a desire to represent the mysterious and autonomous powers of nature leads Pope to the very verge of describing a vitalist vision of an independently self-generating natural world. The persistent mystery of nature's powers of vitality lured and frustrated men of science, too, who were often forced, in their attempts to name them, to coin new terminology, to rehabilitate the 'occult' qualities of the ancient world which Newton had derided, or to speak with an ambiguity, reticence or circumspection (such as that of the Scottish physiologist John Hunter) that only generated further debate.

The language of vitality, whether in scientific or literary contexts, could thus often be far from precise, but its periodic ambiguity could be productively suggestive – a fact which in turn may explain something of its hold over the collective eighteenth-century imagination. This ambiguity is itself telling, pointing to the presence and fascination of unanswered, perhaps unanswerable, questions which extended their own pull in an age of enquiry. However it is read, for instance, Pope's *Essay on Man* demonstrates a persistent early eighteenth-century uncertainty about the precise relationship between God and nature, between life's final cause and nature's own material powers of self-generation, which Newtonian science, expounded with whatever poetic skill, could not fully dispatch. Indeed, Newton, who had grappled with the possibility of non-mechanical forces in matter, was himself the source of some of the questions directly related to vitalism. In the General Scholium added to the second edition of his *Principia* in 1713, and in the Queries added to successive editions of his *Opticks*, he supplemented his earlier work's focus on gravity to postulate the existence of shorter-range attractive forces, sustained by a mysterious aether, to explain a range of chemical phenomena. This addition of force to matter proved suggestive, even whilst a series of questions – were such forces mechanical, material, immaterial? – remained. The nature of matter, and the possibility of activity in matter, became a fundamental debate of the age, with ramifications for many forms of scientific, philosophical and theological enquiry. One of its consequences was to enable physiologists to begin to broaden or break down a mechanism which had dominated their field since the late seventeenth century, so that the iatromechanism which had founded knowledge of human and animal bodies was supplemented by more complex models, where non-mechanical forces often found a place.[7] The formulation of new concepts of matter, which encompassed notions of force, power, change and dynamism, as well as the sceptical critique of the epistemological bases of mechanical natural philosophy, thus led directly to the emergence of new vitalist theories of living bodies.

The nature of matter, and the possibility of activity in matter, came under discussion throughout the eighteenth century not only in a natural philosophical context, but in philosophical circles too. It was still a controversial issue (because of its potentially irreligious implications) for the Edinburgh Philosophical Society in 1754, where David Hume, Lord Kames and John Stewart, Edinburgh Professor of Natural Philosophy, clashed on the topic.[8] Locke's suggestion, in his *Essay Concerning Human Understanding* (1690), that matter might somehow be capable of thought, is another important landmark in this debate, and I return to Locke in the second half of this Introduction.[9] Whilst the purpose of this study is to focus on vitalism, and especially vitalist physiology, within Britain, meanwhile, it is worth noting that these debates and developments took place in a European context. The formulation of a vitalist physiology in Scotland, for instance, to which I give a particular focus, took place both in reaction to and in concert with Continental traditions. The teaching of medicine and the development of physiological theory were already practices with strong links to Europe: the Edinburgh Medical School was modelled on the university at Leiden, where the iatromechanism of the medical teacher Hermann Boerhaave reigned; Robert Whytt, Edinburgh's leading vitalist physiologist, studied at Paris, Leiden and Rheims as well as at the Scottish capital, corresponded with doctors across Europe and engaged in extended debate with the Swiss-born Albrecht von Haller. Montpellier, Paris and Göttingen were other important centres for the development of vitalist theory, and the Frenchmen Boissier de Sauvages, Théophile Bordeu and Paul-Joseph Barthez, and the German Johann Friedrich Blumenbach, were equally significant figures in the development of vitalism on the Continent.[10]

Within Britain, new vitalist accounts of nature emerged most notably with developments in Scottish physiology from the 1740s, within a decade or so of Pope's poem.[11] The Edinburgh Medical School was a leading educational centre throughout most of the eighteenth century, far outclassing Oxford or Cambridge, and only surpassed by the rise of Paris at the century's end. One reason for its pre-eminence was the opportunity for clinical teaching in Edinburgh hospitals, but another was the existence of the critical, clubbable atmosphere of enquiry of the Scottish Enlightenment.[12] It was in such conditions that the iatromechanism propounded by Boerhaave, teacher at Leiden, to which Edinburgh had strong links, first came under attack: not by the professors but in a student medical society, a forum which encouraged the interrogation of established opinion. William Cullen, who later became an influential medical teacher at Edinburgh, as well as a life-long friend of Adam Smith and a supporter of David Hume, was a founding member of this society; Robert Whytt, who was to become another prominent member of the Edinburgh faculty, was also strongly associated with it. Whytt became a leading figure in the development of vitalism, formulating an 'animal œconomy' in which the body was governed and regulated by an all-pervasive life-force, a 'sentient principle' co-extensive with the body and present from birth to death. His vitalist physiology proposed that such a life principle constituted an animated force which united body and mind, co-ordinated the vital functions of essential organs and ensured the unconscious, immediate, self-preserving responses of the body to external stimuli. It was a model in which the traditional role of the mind as the over-seeing and conscious governing function was displaced by a physiology of co-ordinated, unconscious and independent bodily regulation: a displacement of the thinking mind by the body's own powers, with a strong emphasis on the contribution of the nervous system. Although Whytt used the word 'soul', his physiology was not to be confused with the animism of Georg Stahl: the involuntary bodily activity he described was not controlled by a rational centre. The popular and influential teachings of Cullen similarly emphasised the body's 'nervous power' as a dominant life-force. With the establishment of Edinburgh as a prominent centre for medical education, Whytt's ideas and those of like-minded colleagues and students were to have a significant influence on developments both within medical theory and practice and beyond it, in adjacent and overlapping areas of natural philosophy.[13]

Such vitalist doctrines, with their dramatic remodelling of the functioning of the human body and its relationship with the reasoning powers of the mind, soon made their effects felt in fields beyond the sciences of life. In some ways itself a product of the interdisciplinary enquiry which character-ised the Scottish Enlightenment – in which, for instance, a concept such as 'sympathy' operated within both moral philosophy and physiology – Whytt's ideas about the animal œconomy resonated with new formulations of a non-interventionalist, self-preserving system of political economy being produced by his contemporary Adam Smith, and with developing notions of moral response and sympathetic sensitivity in moral philosophy. As Chapter 3 describes, the Edinburgh Philosophy Society, attended by Whytt, Smith, David Hume and other notable figures, provided one forum in which such cross-fertilisation of concepts and ideas between different forms of enquiry may have taken place, and vitalist ideas spread beyond Edinburgh in other ways. The influential physician and anatomist William Hunter propounded an Edinburgh-acquired physiology in his well-attended London anatomy lec-tures, and the far-reaching experimental and theoretical work of his brother, John Hunter, developed the vitalist understanding of the body as a system-atic, self-communicating organism, by emphasising its automatic powers of healing and locating a force of 'vitality' in the blood. At the same time, others speculated that the 'vital principle' might be equivalent or similar to electricity, or identified vitality with the newly discovered oxygen. Although (Joseph Priestley most obviously excluded) these figures largely steered clear of speculation regarding the religious or other implications of their work, their new accounts of the functioning of the body were certainly suggestive in a number of ways. Potentially, by displacing the governing role of reason, vitalist physiology rewrote the understanding of the human subject, and opened up ways for subconscious or unconscious acts to be contemplated or imagined. Theories of vital energies in nature challenged a traditional account of creation directly controlled by God's will, and offered instead a possible perception of an independent, materialist and self-determining nature; and implicitly political metaphors of 'ungoverned' self-regulation invited obvious application to the world of human affairs. By the time of the 'vitalist controversy' of 1814–19, a dispute between John Abernethy and William Lawrence over the vitalist or materialist interpretation of John Hunter's work, vitalism's ability to attract and focus religious and political controversy gained widespread public recognition, but its connection to vari-ous forms of radical political and scientific thought was firmly established well before then. Vitalist theories of physiology, nature and matter were suggestive for political thinkers, theorists and philosophers keen to rethink and reform man, nature, society and government, and in the revolutionary debates of the 1790s, a language of animation, vitality and liberty was taken up by writers of all kinds – to the extent that invocations of innate 'energies' in man or nature became both a byword of Jacobinism and, perhaps unwit-tingly, repeated by their loyalist opponents. By the end of the century, vital-ism, rightly or wrongly, had become identified with a controversial and fully fledged materialism, whose political, theological and scientific implications, in distancing God from nature, and insisting on the autonomous powers of matter, were fully recognised by both proponents and critics.

Despite its evident importance to the eighteenth-century scientific, religious, philosophical and cultural contexts already outlined, vitalism has primarily been studied to date in the context of the intersection of Romantic science and Romantic literature and culture. Sharon Ruston's *Shelley and Vitality*, for example, explores the connections between the writings of Percy Bysshe Shelley and the Abernethy–Lawrence vitality debate, whilst Nicholas Roe's *The Politics of Nature* focuses on connections between vitalism and Wordsworth in the 1790s.[14] Alan Richardson's study of conceptions of the mind–body relationship in this period, *British Romanticism and the Science of the Mind*, also expounds on material connected to vitalism within a context defined by Romanticism, and this field as a whole benefits from a recent burgeoning of interest in Romantic science itself, the secondary literature on which is too extensive to consider here in detail.[15] Such work clearly represents a valuable extension to, and historicisation of, traditional critical accounts of Romantic organicism, but to limit a study of the connections between vitalism and literary culture solely to the Romantic period represents an arbitrary foreshortening, especially given the origins of much so-called Romantic science in the vitalist natural philosophy and physiology of the mid-eighteenth century. The transformations in understandings of the nature of man, the body, nature, matter and life, which critics locate at the end of the eighteenth century and the beginning of the nineteenth, in fact have significant beginnings decades earlier. Thus whilst, in discussions of Mary Shelley's *Frankenstein*, it has become a critical commonplace to invoke various forms of contemporary experimentation on the animation of bodies, critics often fail to concede that such enquiry into the nature of vital forces and living principles, an associated scientific theorising about the nature of life and disturbing experimentation on live or near-dead animal bodies, can be traced back at least to the 1740s. (As I suggest in Chapter 6, Shelley's backdating the action of her novel to the last decade of the eighteenth century – 20 years before its publication – may be especially significant in this context.) Equally, whilst for Ruston and Roe, figures of animation in Percy Shelley or Wordsworth signal Romantic poetry's response to vitalism, they fail to note the significant history of a language of animation, and an understanding of poetry as a force which, in Samuel Johnson's words, 'animates matter', in eighteenth-century poetry and poetics.[16] To do justice to a complex history of the representation of, and investigation into, animation in this period, and its effects in literature, culture and politics, it would appear that the boundary between 'the eighteenth century' and 'Romanticism' – which itself only makes sense in terms of an outdated literary history – must continue to be renegotiated.

The more extended account of vitalism which is offered here thus demands that we rethink more carefully precisely what is being named when we invoke Romanticism and its difference from an earlier eighteenth-century period – or when we found such a difference on a perceivedly Romantic concern with vital nature. The rise of contextual historicisation in literary studies has not always been allowed to threaten existing demarcations of literary period, and this has led at times to what can appear to be an overly neat equation between literary Romanticism and Romantic science, both conveniently demarcated from the literature and science of an earlier period. This appears to be implied in Ruston's assertion, for instance, that vitality's 'new way of thinking of life is bound up with a distinct shift in the politics, literature and attitude which constitutes that loose web of ties we call Romanticism', or that the 1814–19 vitality debate 'can be seen as promoting the tenets of Romanticism against outmoded eighteenth-century ideals'.[17] But the very vocabulary of life which she identifies as 'contemporary' for Percy Shelley – including the terms 'sensibility' and 'irritability' – was central to a prominent debate between Whytt and Haller 60 years earlier.[18] What does it mean for our understanding of literary Romanticism if its 'contemporary' language of vital life in fact precedes its own moment by a good half-century?

Although a study of specifically eighteenth-century vitalism, the longer historical perspective on vitalism offered by this book, holds out the promise of a sharper understanding of the distinctiveness of both 'Romantic science' and Romantic literary writing. It suggests that, instead of founding a distinction between Romanticism and what preceded it on an outmoded account of eighteenth-century mechanism and Romantic organicism, the continuities between the sciences of life in the second half of the eighteenth century and in the Romantic period must be properly recognised. To understand Romantic writers as inheritors of this tradition can only enhance our sense of what they did with it. Whilst vitality in an eighteenth-century context, on the one hand, is a theoretical formulation deployed in relation to specific scientific questions, about life, matter, bodies and so forth, and on the other, is a suggestive language of animation informing political, economic and literary writing, Romanticism's totalising, unifying urges transform a technical postulation of vitality within experimental and observational science into a generalised creative force existing throughout a dynamic universe: in organic nature, in creative genius, in the poetic mind and transcendent spirit. Such a Romantic vision of a vital organic universe, harmonised and unified, was central to its resistance of the fracturing of knowledge into distinct, technical disciplines, so that vital life becomes a central, transcendent metaphor, rather than a precisely defined field of enquiry. The attention paid to what we now call science by literary writers such as Coleridge or Shelley was thus part of their attempt to heal divisions in knowledge through a transcendent poetic speech or vision; an attempt, at one level, to bridge the modern 'two cultures' of science and art which current literary criticism, by carefully redescribing a scientific context for

Romantic poetry, risks unintentionally reinvoking, despite its disavowals. Coleridge's claim that he attended Humphry Davy's lectures on chemistry to 'increase my stock of metaphors', although it reads like a joke, might thus be understood entirely seriously, and as marking a conscious Romantic sense of the instrumental relation of science to literature through which poetry's overriding visions might be achieved.[19]

   This book attempts to counter the occlusions of previous work by placing vitalism in a wider historical and geographical frame. Rather than understanding it as a radical Romantic science, whose 'politics of nature' chimed with youthful poetic idealisms, this book argues that vitalism needs to be understood in the context of intersecting eighteenth-century enquiries in a range of discourses into the nature of life, matter and the human subject. Ironically for a science which was later to be denounced as 'materialist', this book argues that eighteenth-century vitalism is associated with the attempt, in scientific and other discourses of the period, to move beyond limited mechanistic formulations on such topics.[20] It sees vitalism as a response to the inability of a mechanically oriented natural philosophy, inherited from the late seventeenth century, to understand the specificity of living systems, or the origins and nature of life, or the operation of the human body, and as a response too to interconnected debates about the relationship of matter to 'spirit' and to thought which could be traced back to Newton and Locke. Vitalism's presence, as this book will argue, in eighteenth-century literary culture and beyond, marks that period's negotiation, in science, philosophy, religion and literature, with the limits of material notions of the human subject, and with the limitations of materialism itself. In this, it offers a new way of approaching the period's well-documented fascination with the non-material – with its notions of animation, transport, vitality and effusion, concepts central to eighteenth-century constructions of sensibility, the gothic and the sublime, and repeatedly reworked in numerous writings on aesthetics and other forms of affectual experience. This book understands vitalism, in this sense, as marking the renegotiation of the relationship between the material and the immaterial, the bodily and the emotions, the subject as embodied and as more than 'body', which so characterised much philosophy, religion and literature of the period.

   These sweeping concerns are focused in the first part of the book through a more particular one: the intersection between vitalism and eighteenth-century understandings of the human subject. Roy Porter and George Rousseau have noted how this period marked the formulation of 'new scientific, secular concepts of personality and identity', an observation echoed by other scholars elsewhere, and considerations of the nature and character of the 'self' in the eighteenth century have occupied many.[21] Vitalism contributes to a multi-faceted debate about the constitution of the human subject in this period by intervening in the account of the relationship between mind and body which was central to understandings of the human person at this time. The eighteenth century inherited from Descartes a model of the human subject broadly understood through a mind–body division: a reasoning, rational mind was thought to control the material body, a model which usefully corresponded to religious conceptions of the human as uniting earthly matter and divine reason. The finer details of such a model, and its division between the immaterial and the material, however, were the subject of heated and often controversial discussion, especially in philosophical and religious circles where, as already mentioned, Locke's speculation that matter might be capable of thought (or that thought might in some way be a material process), resulted in decades of dispute. The ambiguity of the relationship between mind and body, thought and matter, manifested itself in social and cultural anxieties too. Dennis Todd has shown how contemporary responses to disabled, or so-called 'monstrous' births, expressed the fear that the imagination could transgress the barrier between mind and body, a fear he traces in the writings of Pope and Swift, and which could still be seen resonating in the birth of Sibella's stillborn child in Eliza Fenwick's *Secresy* (1795), discussed in Chapter 6.[22] Indeed, the developing theorisation of the imagination at this time might itself be attributed at least in part to a fascination with the productive ambiguity of the mind's relationship to the body. The contribution of vitalist physiology, with its supposition of a vital principle directing essential bodily functions, reflexes and communications, was to challenge the centrality and importance of the mind's ruling function, disrupt a traditional assessment of the relationship between mind and body, and insist on the greater autonomy, unconsciousness and immediacy of the body's self-rule. Against a model of the body as the passive recipient of commands from the mind, it posited an animated body whose forms of unconscious regulation and control nevertheless promoted its best interests and ensured its self-preservation. It thus re-posed the relationship of the subject to the body by emphasising the limits of the mind's control over the body and the autonomy of the body as a self-directing, self-controlling entity, and more broadly by offering a model of the 'self' as animated, vital, fluid and in flux, rather than under the rational, conscious and regulated control of a reasoning mind. Its relevance to the emergent culture of sensibility, discussed in Chapter 2, where such a depiction of the self operated, is thus clearly evident.

This book is not concerned with offering a narrative of such changing accounts of mind and body at this time, however, although a more detailed discussion of this is given in the second half of this Introduction, as a necessary context for the book's key concerns. Rather, it is concerned with tracing the presence, influence and effects of vitalism within certain crucial moments of the 'long' eighteenth century, especially in relation to representations of the human subject, and the ways such formulations inform the discourses of sensibility and political economy, and, in Parts II and III, the radical politics and culture of the 1790s. Part I, 'Writing the Body in the

Scottish Enlightenment', considers the conception of the human subject, mind and body, in the Scottish Enlightenment, chosen both as the centre for a wide-ranging and influential enquiry into the nature of 'man', in discourses including moral and natural philosophy, and as the point of emergence, within Britain, of a fully formulated vitalist physiology. Chapter 1 considers the representation of the human subject and the relationship of mind to body in four Scottish case studies: in the Scottish 'science of man', and in the poetry of Scottish poets John Arbuthnot, John Armstrong and Thomas Blacklock. It discusses a context of fertile cross-dissemination between natural philosophy and literary and philosophical culture in Scotland at this time, a context in which contiguous debates and discussions about the nature of 'man' ensured the speedy assimilation of new scientific and philosophical ideas. Chapter 2 explores the relationship between mind and body in the culture of sensibility, for which Adam Smith's *The Theory of Moral Sentiments* (1759) was a foundational text, and traces how Smith's theorisation of sympathy is both enabled and constrained by the question of its relation to the material body. Mary Wollstonecraft's later accounts of the generation of sympathy are also considered. The final chapter in Part I discusses the representation of the body in economic writings by David Hume, Smith and others, and sees competing mechanistic and vitalistic discourses being used to construct a labouring subject for whom work is both a natural and a necessary endeavour. It also explores the connection between Smith's *Wealth of Nations* and the vitalist physiology being discussed in Edinburgh from the mid-century, to argue that Smith counters contemporary mercantilist economic policies by suggesting that, like the 'vitalist' body, the economic 'body' can best ensure its own health, and pursue its best interests, if left to its own self-regulation.

Parts II and III of the book move on, both in time and geographically, to trace the connections between vitalism and political debate and literary production at the end of the century. Part II, 'Enlightenment in the 1790s: The Scottish Legacy', traces the influence of Scottish moral and natural philosophy on late eighteenth-century radical political writing, focusing on the work of radical campaigner and lecturer John Thelwall. In particular, it traces how a Scottish 'vitalist' physiology, disseminated in London through John Hunter's anatomy lectures, contributes to the emergence of a materialist conception of nature as self-governing and self-animating: an account which threatened established notions of divine power over nature, and whose model of a system operating successfully and 'naturally', whilst independent from external governance, was clearly politically suggestive. For Thelwall, such a theory of nature provides a productive parallel for politically radical accounts of society, and opens a rich metaphorical store to be exploited in his writings. Such concerns were suggestive beyond scientific and philosophical circles, however, and Part III of the book, 'Vitalism, Animation, Culture', explores the deployment of figures of animation, vitality and consciousness in the literature of the period. The natural description poetry of Edinburgh-educated Erasmus Darwin (*The Loves of the Plants*, 1789, and *The Economy of Vegetation*, 1791) offers a poetic exposition of an animated and conscious natural world, whose unconstrained pursuit of its 'loves' hinted at similar possibilities for human behaviour; his posthumously published *The Temple of Nature, or The Origin of Society* (1803) extended such parallels between natural and human worlds by describing the development of human society as an extension of the evolution of life itself. Recurring metaphors of animation, life and death, meanwhile, are especially notable in the work of women writers of the 1790s, including Mary Wollstonecraft, Mary Ann Radcliffe and Eliza Fenwick: the valuation of 'life' and animation in their work operates as a powerful critique of social forces and institutions which are in turn figured as oppressive and deathly. Mary Robinson's rejection of discrimination between man and woman until 'the sex of vital animation ... be ascertained' demonstrates too how a discourse around the scientific nature of animated life informed contemporary debates about sexual difference. Evidently informing a rich cross-section of writing – political, poetic, propagandist, fictional – vitality and animation are, by the final years of the century, more than narrowly scientific or philosophical concerns, but inform the very language and thought by which writers of all kinds address the central concerns of the moment.

Mind, body and vital life in the eighteenth century

Whilst post-Newtonian matter theory provides one context for understanding the emergence of vitalist theories in the eighteenth century, another is provided by debates in philosophy over the relationship between the mind and the body, and between the subject and his or her embodiment. The discussion can be picked up with Locke, whose *Essay Concerning Human Understanding* (1690) mapped the operation (or 'natural history', as he described it) of the mind with a new precision, whilst also renewing questions about the relationship of the mind to the material body.[23] Such questions are particularly focused in a discussion of the nature of 'personal identity' in a chapter added to the *Essay* in 1694, where Locke repeatedly demonstrated how 'identity' – to be understood not, as we might today, as 'personality', 'character' or 'subjectivity', but 'constancy over time' – was distinct from a subject's physical embodiment. '*Self*', Locke asserted, 'is that conscious thinking thing, (whatever Substance, made up of whether Spiritual, or Material, Simple, or Compounded, it matters not) which is sensible, or conscious of Pleasure and Pain, capable of Happiness or Misery, and so is concern'd for it *self*, as far as that consciousness extends.'[24] Locke's refusal to equate selfhood with a physical body, and his openness about the nature of the 'Substance', whether material or immaterial, in which 'Self' exists, is striking, and this was emphasised by his later examples, which undermine a natural tendency to associate the self with a body. Should our little finger become separated from our body, he continues, and in addition our consciousness left our body to join that finger, then "'tis evident the little Finger would be the *Person*, the *same Person*; and *self* then would have nothing to do with the rest of the Body'.[25] Wrong-footing his reader by insisting on the potential redundancy not of the little finger, but of the body, Locke breaks a traditional and automatic association of self with embodiment, and insists on consciousness as the constitutive element of selfhood. At the same time, however, his close association of consciousness with the physical experience of sensation, including the sense of pleasure and pain, and his assurance that 'self' has a 'Substance', reiterates the question of the relation between the consciousness which defines the self, and the substance in which it inheres. Locke has both broken an unconsidered equation between mind and matter, between the thinking self and the body it inhabits, and renewed it, albeit with a new uncertainty about what the nature of body, or 'Substance', might be.

Locke troubles the relationship between self and body – personal identity is not automatically equatable with physical embodiment – but also insists on it, given a necessary relationship between consciousness and its 'Substance'. One feels that a new account of what body or 'Substance' might be is necessary here, and something like it is offered in Locke's consideration of the oak tree. In an example later to be reused by Percy Shelley, the changeability of the oak as it grows from acorn to tree, yet its constant identity – '[a]n Oak, growing from a Plant to a great Tree, and then lopp'd, is still the same Oak' – reiterates the same questions about the relationship between 'Matter' and 'Identity'.[26] The example leads Locke to consider the nature of the oak's 'Substance', which he describes as 'such an Organization of those parts [of the oak] as is fit to receive, and distribute nourishment, so as to continue, and frame the Wood, Bark, and Leaves *etc.* of an Oak, in which consists the vegetable Life'. He continues:

> That being then one Plant, which has such an Organization of Parts in one coherent Body, partaking of one Common Life, it continues to be the same Plant, as long as it partakes of the same Life, though that Life be communicated to new Particles of matter vitally united to the living Plant, in a like continued Organization, conformable to that sort of Plants. For this Organization being at any one instant in any one Collection of *Matter*, is in that particular concrete distinguished from all other, and is that individual Life, which existing constantly from that moment both forwards and backwards in the same continuity of insensibly succeeding Parts united to the living Body of the Plant, it has that Identity, which makes the same Plant, and all the parts of it, parts of the same Plant, during all the time that they exist united in that continued Organization, which is fit to convey that Common Life to all the Parts so united.[27]

The definition of the oak which emerges from this tortuous passage is of an 'organization' of parts united and animated through 'Common Life'; 'organization' was to be a central term in eighteenth-century physiology. Its identity depends not on constancy of matter, for its parts are constantly changing, but on constancy of animation, and this in turn shifts the emphasis, in considering a tree's materiality, from the shape or physical character of its parts, to their ability to partake in that vital 'Life'. So far as the 'identity' of an oak is dependant on its 'Substance', that substance needs to be understood less in terms of physical constancy or form, more in terms of its ability to continue and convey the essential quality of 'Life' which ensures the tree's identity, an ability more significant than questions of 'Substance's' material or immaterial status. What is most pertinent about the character of 'Substance' is not its physical form, its material or 'spiritual' being, but its essential ability to support life.

The same move, to equate identity not with body but with life, is made in Locke's discussion of mankind. Here his account can be seen to respond to a context in which, given the influence of natural philosophy's mechanical metaphors on empirical philosophy and contemporary physiology, it had become necessary to describe how man's identity was more than merely matter – as well as to reconcile knowledge of man's corporal being with some account of how matter is animated. At stake in Locke's discussion is the distinction between the organised matter of 'man' from that of machines, a distinction which he secures by emphasising the animating spark of life itself:

> [W]hat is a Watch? 'Tis plain 'tis nothing but a fit Organization, or Construction of Parts, to a certain end, which, when a sufficient force is added to it, it is capable to attain. If we would suppose this Machine one continued Body, all whose organized Parts were repair'd, increas'd or diminish'd, by a constant Addition or Separation of insensible Parts, with one Common Life, we should have something very much like the Body of an Animal, with this difference, That in an Animal the fitness of the Organization, and the Motion wherein Life consists, begin together, the Motion coming from within; but in Machines the force, coming sensibly from without, is often away, when the Organ is in order, and well fitted to receive it.[28]

Man's 'identity', Locke continues, consists in 'nothing but a participation of the same continued Life, by constantly fleeting Particles of Matter, in succession vitally united to the same organized Body'.[29] As with the oak tree, the identification of 'one common Life' here defines man's identity as more than merely material, usefully and necessarily distinguishing 'man' from a machine. Locke's characterisation of 'identity' through the fleeting characteristics of animated life begins to unfold a new language for the human subject, in which immateriality, flux, changeability and inconstancy – united by a 'common Life' - might be central terms. This is especially evident in the formulation, by a critical Joseph Butler, of Anthony Collins' extension of Lockean ideas on identity. Some have gone to particular lengths, Butler commented, to suggest that

> personality is not a permanent, but a transient thing: that it lives and dies, begins and ends continually: that no one can any more remain one and the same person two moments together, than two successive moments can be one and the same moment: that our substance is indeed continually changing.[30]

Personal identity here – marked by transience, mortality, changeability, dynamism and fluidity – has begun to take on the characteristics of 'life' itself.

Locke's emphasis on 'Life' as the distinguishing characteristic between men and machines, and as an integral part of 'identity', placed an inevitable weight on knowing quite what 'life' might be, and thus connected with a series of interrelated debates central to eighteenth-century vitalism: about the nature, origin and causation of life's animating forces; their relation to the 'matter' which they enliven; whether they are immaterial spirits; whether organised matter alone is capable of generating life; or whether matter has some additional quality, electric, nervous or otherwise, which constitutes 'life'. Such debates continued long after Locke, to the end of the century and beyond. Vitalist physiology, which attempted to identify, or at least postulate, the 'vital' spark animating life forms, was evidently responding to such questions, but so too were other forms of natural philosophy, including matter theory and (given the possible roles of electricity or oxygen) chemistry and pneumatics. Such enquiries intersected too with developing physiological research on the mind–body relationship, another area of some uncertainty and dispute. Locke's own teacher, Thomas Willis, who published the landmark *Cerebri Anatome* in 1664, had himself defined such a field by initiating serious and detailed research into the brain, the senses and the nervous system. His work extended knowledge of the complexity of man's physical being, and pointed to the permeability of the corporeal and the psychological, with its attention to the connections between the impressions of the senses and the ideas of the mind – a relationship already emphasised in the empirical philosophy of the day. In this way, the body could be understood both as a material entity, with functions and processes which could, and frequently were, mapped according to the laws of the physical world, and as something which engaged with, and posed the question of its relationship to, the immaterial or mental.[31] The increasing elaboration, over the course of the eighteenth century, of the body-wide nervous system of reflex, response and communication, thus demonstrated the existence of nervous, sensory and muscular machinery central to empiricism's picture of man, through which the sensations of the body were collected and communicated to the mind. But it also potentially undermined empiricism's account of the centrality of the mind in man's engagement with the world, as an alternative picture, of an autonomous body whose responses and reactions occurred independently from any overseeing mental function, also began to suggest itself.

Such work on the nervous system, whilst at one level pushing back the frontiers of knowledge of the physical body, at another simply restated, and remained defined by, the enigma of the relationship between mind and body. For Descartes, that relationship had been determined by the pineal gland, but for later thinkers things were rather less clear. In Hume's *Enquiry Concerning Human Understanding*, the evident 'mystery' of the relationship between the will and bodily action served as a crucial demonstration of the unknowability of the 'secret connexion' between cause and effect:

> We learn the influence of our will from experience alone. And experience only teaches us, how one event constantly follows another; without instructing us in the secret connexion, which binds them together, and renders them inseparable ... We learn from anatomy, that the immediate object of power in voluntary motion, is not the member itself which is moved, but certain muscles, and nerves, and animal spirits, and perhaps, something still more minute and more unknown, through which the motion is successively propagated, ere it reach the member itself whose motion is the immediate object of volition. Can there be a more certain proof, that the power, by which this whole operation is performed, so far from being directly and fully known by an inward sentiment or consciousness, is, to the last degree, mysterious and unintelligible?[32]

Hume's sceptical attack on the knowability of cause and effect turns to the ambiguities of the body for demonstration – although, at the same time, the rhetoric of his argument almost leads him into offering some kind of supposition of causal relation: 'something still more minute and unknown'. Despite the mystery, 'something', he cannot help wondering, must be at work communicating between the will and the body, but whatever it is, it lies beyond our senses, and hence beyond empirical discovery. A similar move, on the same subject of the relationship between mind and body, marks the conclusion of the General Scholium which Newton added to his *Principia* in 1713, and its meditations on the limits of inductive philosophical method. Addressing precisely the same question as Hume, Newton confidently suggests the existence of 'a certain most subtle Spirit' by which 'the members of animal bodies move at the command of the will, namely, by the vibrations of this Spirit, mutually propagated along the solid filaments of the nerves, from the outward organs of sense to the brain, and from the brain into the muscles', before remembering the lack of a 'sufficiency of experiments' needed for 'an accurate determination and demonstration of the laws by which this electric and elastic Spirit operates'.[33] Only his previous paragraph, after all, had demanded the expulsion of hypothesis from experimental philosophy. Finally, the cause of bodily motion, and the associated possibility of some obscure animating function, coincides again with the question, and limits, of empirical philosophical methods in Newton's *Opticks* (1704). There, the ancients' explanation of the motion of bodies by 'occult qualities' is condemned as putting 'a stop to the Improvement of natural Philosophy': to invoke occult qualities – mere suppositions, unproven imaginings – 'is to tell us nothing'.[34] For both Newton and Hume, the particular question of the mind's relation to the body, and the possible animating forces which link them, is repeatedly bound up with, and impeded by, the niceties of natural philosophical method. To proceed beyond this impasse, vitalists recognised that their science was necessarily hypothetical, an attempt to sketch what might exist, rather than to demonstrate what surely did. If the vital principles of man's animation could not be known, they must nevertheless somehow be postulated.[35]

The vitalism associated with the Edinburgh Medical School from midcentury can be seen to have taken precisely such a step into supposition, offering in effect a hypothetical account of the relationship between mind and body (and hence of the nature of animation and the life of bodies), which was nevertheless hugely influential in terms of both physiological and anatomical theory, and in medical practice. Robert Whytt, professor in the medical school from 1747 and a leading figure in this context, to whom we return in Chapter 3, certainly understood himself to be a good Newtonian – a proponent of inductive methodology in natural philosophy. Writing in the preface to his *Essay on the Vital and Other Involuntary Motions of Animals* (1751) – a work which addressed, as had Hume and Newton, precisely the question of the influence of the will on the body, and the cause of bodily motion – Whytt repeated the usual methodological assurances which might be expected of one working in the tradition of Newtonian method: he has been 'careful not to indulge his fancy, in wantonly framing *hypotheses*, but has rather endeavoured to proceed upon the surer foundations of experiment and observation'.[36] Whytt's assurances to his reader that his 'just account' of vital motions is derived 'rightly from their true source', however, sits oddly with his recognition that 'nature has so closely concealed many of her operations, that they often elude the united efforts of genius, industry, and experiment', and his acknowledgement of the value of the 'unguided imagination' in 'hitting' upon the truth, might seem to betray a looser understanding of the need for inductive demonstration than more orthodox scientists might have wished.[37] Certainly, Whytt's assertion of a 'sentient principle', or a 'living principle capable of generating motion', was not an experimental deduction but a theoretical postulation, an imaginative sketching of a form of communication or connection existing in the body designed to fill the gaps in what could be known more definitely about the 'animal œconomy'. But if the exact nature of that 'sentient principle' was in the final analysis rather hazy, its supposition nevertheless enabled physiologists to move beyond the obstacle of the limits of empirical demonstration, as experienced by Hume and Newton, and see – or perhaps imagine – a more complete account of bodily animation: a comprehensive, if in some ways hypothetical, vision of the self-animation of living matter.

Vitalism offered a solution to the debate about the nature of the mind/ body relation by dethroning the mind from its assumed role of reasoned governance of the body, and by envisioning, instead, a body capable of automatic and autonomous, if unconscious and instinctual, self-direction and self-preservation. In place of the directive responsibilities of mind, and reason, over a passive and submissive body, this new version of the 'animal œconomy' offered a new explanation of the powers of what was quite literally a 'body politic', to replace 'king reason' with nerves, senses and unconscious instinct. 'Life' itself began to look rather different: no longer a physical entity passively carrying out the orders of reason, but a fluid, constant, dynamic, changeable and ultimately elusive force, existing and communicating throughout a vitally animated body. This book traces, over the course of the following chapters, the presence and effects of such a new model of 'life' in literary, cultural and political contexts, but the transformation in the conception of the nature of human life can perhaps immediately be seen in a more practical context. In 1776, the same year his patient Adam Smith published *The Wealth of Nations*, and, as Chapter 3 describes, exhorted the virtues of a 'vital' nature's self-healing capacities to the dying David Hume, William Cullen, Edinburgh colleague of Robert Whytt and a prominent medical teacher, theorist and practitioner in his own right, published *A Letter to Lord Cathcart, President of the Board of Police in Scotland, Concerning the Recovery of Persons Drowned and Seemingly Dead*. Responding to an invitation from Lord Cathcart, Cullen denounces the very limited endeavours then taken to recover drowned persons, and outlines at length a series of measures aiming to restore the 'vital principle' to such casualties: ranging from warming and chaffing the body to restore its 'vital heat', immersing it in a warm bath, applying liquid stimulants, blowing tobacco smoke into the 'intestines' or air into the lungs, and even opening the veins. Recognising that such actions might take some time before producing an effect, Cullen recommends that they are continued for 'several hours', or until, if no sign of life appear, the 'symptoms of death ... go on constantly encreasing [*sic*]'.[38] Remarkable for its author's faith in the possibility of restoring life even hours after a patient's possible death, the pamphlet's recommendations aim at restoring 'the vital principle': a 'certain condition in the nerves, and muscular fibres', on which the 'living state of animals ... especially depends'.[39] The conception of 'life' which emerges from all this is not of a single animating spark or God-given command, of dull 'clay' all too briefly and fleetingly animated, but as a condition of the body – of the mysterious 'nerves and fibres' – whose physical condition can be manipulated to restore life even several hours after apparent death, as though residual, unconscious, instinctive powers might even at that late stage be spurred into self-preservatory action. Equally, the sheer variety of interventions and recovery procedures recommended by Cullen suggests both the importance of the 'vital forces' which produce and preserve life, and their ultimately mysterious, unknowable nature.

Cullen's notion of our 'vitality', or the 'vital principle' on which life depends, might be vague – 'a certain condition' in the nerves and fibres – but it nevertheless attempts to sketch, at a physiological level, a sense of a life-force at once transient, mortal and capable of being lost or broken, yet also sufficiently malleable or latently responsive to be recovered even beyond apparent death. These characteristics replay, within physiological discourse, the same sense of something living vitally on, beyond apparent breaks and fissures in a mere physical state, as Locke's accounts of personal identity, discussed earlier, appeared to be reaching for. Cullen's medical prescriptions and practices share with Locke's philosophical analysis of the mysterious nature of identity in living beings the same sense of a body's elusive but central 'vital unity' – as well as the same inability to define it; for both, vitality is a matter of necessary assumption.[40] Certainly, both insist on a vital connection which maintains life even across the differences of physical change and apparent death, and even in the face of the difficulty of describing quite what that vital 'something' might be. That very difficulty of description, experienced in both scientific and philosophical discourses, and the necessity of a kind of assumption or faith or what might be called a sketchy, hypothetical fiction, might in turn suggest that fictional or imaginative writing may have provided a more productive and amenable space to rehearse these recurring concerns. One response to Locke's account of personal identity, an Oriental fantasia of adultery, treachery, switched identities, vulnerable monarchs and deceitful subjects, certainly continues within the space of literary writing the very preoccupations with life, death, identity and vitality which we have been tracing here. Published in *The Spectator* in 1714, in an explicit response to Locke's chapter on personal identity, the Persian tale describes how Fadlallah, king of Mousel, is tricked by a dervish who has the 'Power of re-animating a dead Body, by flinging my own Soul into it'. Curious to experience such a power himself, the king 'shoots' himself into the body of a deer, only to have his own body usurped by the dervish, who thus takes Fadlallah's kingdom and sleeps with his wife Zemroude, whilst the king (who now transfers himself into the bodies of his wife's nightingale and then lapdog) can only look on in horror. It is only when the dervish, comforting Zemroude for the death of her nightingale, reanimates it by 'shooting' himself into it, that the king is able to regain his own body, kill the dervish in the reanimated nightingale and regain his powers, only for Zemroude to die with grief at her 'innocent Adultery'.

Playing with a sense of identities both confused and transferable, and curiously persistent (the king is still the king, even whilst in the body of the lapdog), of 'selfhood' as distinct from, yet most properly allied to, physical embodiment, of mysterious powers over life and death, and of reanimation of the dead, the tale pursues the same questions as Locke and Cullen had addressed about the nature and relation of mind and body, identity, life and death. Within the exotic space of the Oriental tale, it anxiously rehearses what these new possibilities might entail, and depicts treachery, deceit, and political and marital betrayal as the consequences of newly 'unfixed' notions of life, identity and selfhood, only balancing such fears with the reassuring constancy of Fadlallah and Zemroude's mutual love. Zemroude's death functions as the ultimate sign of that love, but its power to console the reader, if not Fadlallah, is undermined by the ambiguity of what that death might signify about the relation of self and body, the very subject of the entire story. On the one hand, Zemroude's death reasserts the very association of the self with a specific body which Locke's philosophy, and the tale as a whole, had thrown into doubt: she dies precisely because her body, and therefore her self, has been violated by intimacy with a man other than her husband. What is described as her 'extream [*sic*] Delicacy' is thus an inability to dissociate self and body in the ways in which the dervish is so treacherously fluent. But on the other hand, Zemroude's death also renders that inflexible association of body and self as fatal: her certainty that her relations with the dervish were adulterous insists on recognising a treacherous 'self' within or beyond the king's body with which she otherwise innocently sleeps. Zemroude's faith in the identity of body and self – both hers, and her husband's and lover's – is thus at once reassuring and tragic, figured as a marital loyalty which is both admirably excessive and ultimately fatal.[41]

This tale offers just one instance of the ways in which the preoccupations, fascinations and debates on vitalist themes in eighteenth-century philosophy and natural philosophy are repeated and reworked in literary culture. Already, at the beginning of the century, philosophy and fiction are both engaging with the complexities of new notions of the self, identity and life, which the debates described above were rendering newly imaginable. How such debates informed subsequent thinking and writing, not only about human identity and subjecthood, but in politics, economics, poetry, fiction and moral philosophy, both within and beyond literary culture for the remainder of the century, is the subject of the following chapters.

Part I
Writing the Body in the Scottish Enlightenment

1
Forms of Enlightenment: Embodied Beings in Eighteenth-Century Scotland

In 1734, John Arbuthnot, Scottish physician, Royal Society member, Scriblerian wit and close friend of Alexander Pope, published *Know Yourself,* a 138-line philosophical poem. Strikingly different from his other satirical and scientific works – he made frequent, if often unattributable contributions to the Scriblerian output, and published a number of works on mathematical and medical topics – the poem is suffused with an alert and questioning self-consciousness, and attempts to offer a philosophical and theological answer to the injunction to 'know yourself' imposed by its title. *Know Yourself* identifies a 'double Nature' and 'double Instinct' in mankind, whom it places in an uneasy position between heaven and the beasts: 'Angel enough' to seek happiness, but 'Brute enough to make the Search in vain'. The dominant sense is of mankind's postlapsarian displacement and rootlessness, man's ejection from the 'Bliss' of his 'native Sky' and his painful, weary inhabiting of 'this poor Clod', an experience of dislocation which prompts a continual, uneasy and unrealisable desire. Man is a thoughtless wanderer on the road of his life, unattentive to his eventual end, longing to raise himself on 'Wings of Love and Praise', yet making best progress when his 'wary Footsteps' are guided by 'humble Thoughts'. This constrained and difficult position between heavenly aspiration and our fallen nature is ultimately resolved by divine revelation: the 'sacred Text' which uncovers the 'Secret' of man's 'high Descent' and decodes the 'mystick Tokens' which are 'Marks' of man's 'Birth'. The poem thus discovers an enlightenment of sorts, though a temporary one, and one which only reinstitutes the troubling duality of man's nature, its combination of 'unthinking Clod' and 'heav'nly Fire'. By the end of the poem, this doubleness renders a final paradox: whilst the 'Mysterious Passage' of man's eventual ascent to heaven is 'hid from human Eyes', in 'Soaring you'll sink, and sinking you will rise'. In a poem infused with a sense of homelessness and exile, and where man's fallen nature is a mark of regression, man's homecoming is paradoxically achieved through a reconciliation to his fall.[1]

Although ultimately contained within theological conventionalities, the poem's restless quest for self-understanding expresses the search for knowledge of the human subject characteristic of the Enlightenment age in which Arbuthnot lived. Even the poem's title, *Know Yourself*, a slogan adopted from classical philosophy (and used too by Pope for his own short poem), articulates something of the eighteenth century's fascination with the self. Although for Arbuthnot, seriously ill when the text was published (he was to die the following year), this self-enquiry is resolved – or curtailed – by religious faith, the poem rehearses questions central to contemporary debates over the nature of the self, identity and consciousness, debates he had satirised elsewhere with his Scriblerian colleagues, and which, as we have seen in the Introduction, would become central to vitalist concerns over the course of the century.[2] More particularly, and aptly, given Arbuthnot's medical profession, these questions of self-identity are asked in ways which repeatedly bring mankind's material nature into the frame, and a marked pull towards considerations of human physical embodiment, with all its associated problems, is evident. The poem opens by asking not 'who' but 'what am I?', and proceeds to offer a series of alternative interpretations of man's corporeal being:

> What am I? how produc'd? and for what End?
> Whence drew I Being? To what Period tend?
> Am I th' abandon'd Orphan of blind Chance;
> Dropt by wild Atoms, in disorder'd Dance?
> Or from an endless Chain of Causes wrought?
> And of unthinking Substance, born with Thought?[3]

Arbuthnot is deliberately invoking key questions in eighteenth-century philosophy regarding human identity, origin and purpose, and the controversial materialist answers (the 'disorder'd Dance' of 'wild Atoms') which they might receive. The possibility that man is simply the 'Orphan' of chance counters the philosophical and theological commonplace that mankind is a perfectly designed, divine creation. The 'endless Chain of Causes' questions the notion of God as the first or motivating causes of creation, and the paradox of 'unthinking Substance, born with thought', alludes to the controversial 'thinking matter' debate: the attempt, beginning with Locke and continuing throughout the century, to reconcile a materialist account of man's cognitive functions with a refusal to regard thought as merely a material process.[4] At the heart of all such debates is the problem of reconciling the two aspects of mankind's embodied nature: his physical being and his animated, metaphysical self. Even though these questions are cast in unauthoritative form as anxious questioning by a rootless subject, Arbuthnot is nevertheless audaciously staging a conflict between the possibilities which they raise, and the answers which will ultimately be found in the theological verities which conclude the poem. Although the poem ends by denouncing the Icarus-like 'Wings of vain Philosophy', these determinedly material speculations nevertheless give dangerous space to a godless vision of mankind as no more than – in a phrase Mary Wollstonecraft would use at the end of the century – 'organised dust'.

Arbuthnot's poem demonstrates the extent to which the new science of his age, with its developing empirical knowledge of matter, including human physicality, offered a potential challenge to established theological accounts of mankind. Its fascination with the human subject's material embodiment recurs, as we shall see, in much moral and natural philosophy of the time, informed as they were with the new discoveries and empirical methods of experimental science; and as Arbuthnot's poem shows, a whole new language of physicality, materiality and the bodily spills over into literary writing too. Such writing most often constructs the body in mechanical terms, in line with the dominant mechanical metaphors of early eighteenth-century natural philosophy. It is there, for instance, in Arbuthnot's image of the 'Dull and unconscious' circulation of the blood through the body's 'Pipes', in which the contemporary notion of bodily hydraulics resonates, which modelled vascular and circulatory systems via mechanical laws of fluids – precisely the model of the body which was to be challenged by vitalist physiology within a decade. But such mechanical images of man's physical embodiment often retained a sense of inadequacy or incompletion, as though their possibilities must be weighed against what they fail to represent, or as if an acknowledgement of the mechanisms of human physicality only opens another set of problems in understanding the human subject. Thus in *Know Yourself*, allusions to human corporeality come full circle to reiterate the poem's overriding theological sense of mankind's double nature: although renewing itself with 'transcendent Skill', the body 'Waxes and wastes', combining a divine creativity with earthly decline. In Arbuthnot's poem and beyond it, mechanical metaphors are found to be unable to do justice to a full account of the human subject. Such processes of physical renewal, of 'waxing and wasting', which cannot be explained in mechanical terms, were of course to be central to vitalism's alternative emphasis on the body's internal self-generating powers. Dissatisfaction with the limits of mechanistic accounts of the human body would be a fundamental spur to the growth of vitalism's alternative theories, and its language in turn spills out beyond narrowly scientific contexts to inform other kinds of writing.

The problem of 'personal identity', or the relationship between the body and the self, as discussed by Locke and as central to vitalism's debates, is also present in Arbuthnot's poem. Whilst '[t]he Mansion [is] chang'd, the Tenant still remains': for Arbuthnot, as for Locke, the human subject is a constant tenant in a changing home. Arbuthnot's image casts the 'waxing and wasting' body as emblematic of postlapsarian inconstancy, and contributes to the poem's overall sense of dislocation, but it also points to the same fundamental disjunction between self and body as Locke had explored: the body clearly contains the self, but the self just as clearly exceeds its physical embodiment. As with Locke, the poem's quest into the nature of self both identifies the self with, and dissociates it from, the body – the speaker's body is 'Mine, not Me' – but unlike in Locke, a distanced and troubled identification with the self's material embodiment crystallizes, without resolving, the poem's larger senses of dislocation and homelessness. If in Locke, as we saw in the Introduction, this dilemma is resolved by turning instead to a renewed formulation of the nature of the 'life' which animates the body, in Arbuthnot, the materiality of man's body remains as a more permanent sign of the dissatisfactions and incompletions of human nature; and the duality of man's embodied being is only resolved when the body is left behind in a final ascent into heaven. The poem's repeated figurations of its curiously interstitial and restless sense of what it is to be human all make use of momentarily compelling, but ultimately unsustained, physical powers, whether in the phantasmagoric winged body with which the speaker attempts to 'mount' to heaven, only to 'flag', 'drop' and 'flutter in the Dust'; in the figuring of enlightenment as a form of vision, glimpsed briefly part-way through the poem only to be 'hid from human Eyes' by the end; or in the brutish wants and cravings of the 'two-legg'd Beast'. Human physicality, the powers of the body, are that in which this poem of belief has the least faith. Materiality, for Arbuthnot, does not provide the answer to the questions of human existence which the poem revolves, and its empty shell is finally rejected in favour of a divine life which can only be fully experienced after death.

Arbuthnot's poem encounters many difficulties in its quest to 'know yourself', but they are never allowed to derail the confidently questioning self whose voice and enquiries motivate the text as a whole. The coherence of the poem's speaking persona, even in the face of repeatedly unfolding uncertainties about selfhood, exemplifies the persistence and energy of early to mid-eighteenth-century investigations of human nature which would, by mid-century, produce new vitalist models of the human subject. This explosion of interest in 'man' was characterised by a desire to enquire into all aspects of the nature of the human subject, from the operation of his mind and body, his understanding, imagination and senses, to his anatomy and physiology, his psychology, morality and social behaviour, and his pursuit of wealth, his compassion for others, his aesthetic sense and his literary creativity. As the extensive nature of such a list implies, such concerns were manifested across a range of eighteenth-century writing, in natural and moral philosophy, theology and scientific writing, as well as in poetry and literary productions. At its most explicit and self-conscious, in the various philosophical projects of the Scottish Enlightenment, such a concern to investigate and represent human nature was known, by analogy with natural philosophy's science of nature, as a 'science of man', but, as the example of Arbuthnot suggests, the nature of 'man' was also written through the less formal, but no less influential modes of literary output, and other kinds of cultural activity. This chapter explores the eighteenth-century representation of the human subject, the fate of mechanical accounts of the body, and the problem of human physical embodiment to which vitalist theories responded, in three further contexts: in Scottish Enlightenment moral philosophy; in a poetic treatise on health by John Armstrong; and in the writings of, and various cultural responses to, the blind poet Thomas Blacklock. All, like Arbuthnot, are united by their relationship to the Scottish Enlightenment, where enquiry into the nature of man proceeded on the range of fronts already described, and where, in the decade following the publication of *Know Yourself*, the vitalist theories of the Edinburgh Medical School offered their own solutions to the problems of human embodiment addressed by the poet. All show a fascination with mechanical accounts of the human body, and the variety of ways and contexts in which bodily metaphors can be exploited and deployed. All also demonstrate the kinds of recurring questions about the human subject, and the relationship between material embodiment and human identity, which the literature, culture and philosophy of the Scottish Enlightenment produced. But evident too are the limitations of mechanism as a language of human subjectivity – limitations which help to explain why vitalist physiology emerged in mid-eighteenth-century Scotland as an alternative means of modelling the human body, and, more broadly, of understanding human subjectivity. At times, as we shall see with Armstrong, and even with Arbuthnot, these writers are already reaching towards alternative means of describing a vital, animating force in the human body whose postulation seems necessary to supplement the inadequacies of a narrow, lifeless mechanism.

In the writings explored in the rest of this chapter, the body appears not just as a physical entity, but also as a register for social anxieties and beliefs, an emblem of forms of human ability and weakness, and as a sign of human commonality or difference. The following pages show how writing the body articulates, and attempts to counter, prevalent cultural anxieties, and how it tries to meet social needs. In the poetry of Armstrong, for instance, we see the body as offering compensatory forms of labour and control necessary to invigorate a dangerously indolent leisured class; as providing a regulatory economy to contrast with the excesses of both nature and eighteenth-century consumption; as offering a model of regulation, defeated insurrection and good governance; and the body as enabling, as well as problematising, a socially necessary differentiation between public and private acts, and aiding the production of virtuous subjectivity. The moral dissipation and economic corruption of modern commercial society are written in the somatic terms of bodily exhaustion, but this is countered by an alternative focus on the innate replenishing energies of the body, especially the vital powers of the blood. In Armstrong, the body stands, variously, in metaphoric and metonymic relationship to the human subject, both the sum of the subject and a part of it, both that which is to be administered and regulated by the self, and the 'vessel' or location of subjective experience itself. Such thorough exploitation of the figure of the body suggests how readily new vitalist physiological languages of the body could be received and disseminated at their emergence at mid-century.

This chapter also suggests, however, that whilst a rhetoric of the body was central to early eighteenth-century representations of human subjectivity, such mappings of the human subject worked in conjunction with a conceptualisation of the body which was itself undergoing change. A language of mechanism was giving way to alternative images, even before vitalist physiology began to be formulated. Such slippages are visible even in the medical works of Arbuthnot himself. His *Essay Concerning the Nature of Aliments* (1731), a work in part prompted by the popularity of fellow Scot George Cheyne's *Essay on Health and Long Life* (1724), announces itself to be comprehensible by a reader with as much anatomical knowledge as a butcher, and a 'moderate Skill in Mechanicks', and bodily functions are repeatedly figured in mechanical terms.[5] Thus, the 'Mechanism of Nature' operating in the conversion of food to 'animal substance' involves the mixing of aliment with various animal juices, and the action of 'solid parts ... churning' them together. The precise nature of how food is transformed into other bodily substances, however, remains a mystery which requires an entirely different linguistic register. The 'vital force' of an animal body, Arbuthnot notes, can make milk or blood from food in a way entirely beyond the capabilities of a mere chemist, and such a 'force' can be defined only by its ultimately incomprehensible actions: 'by ... vital Force ... is understood the summ [*sic*] of all those Powers in an Animal Body, which converts its Aliment into Fluids of its own Nature'.[6] Mechanism gives way to, or somehow incorporates, some unknown vital power, whose force is only equalled by its resistance to analysis.

A similar slippage, demonstrating an almost inevitable turn to some 'vital' force, is there too in *Know Yourself*. Man's 'double Nature' is presented as a result of his dual parentage:

> I own a Mother, *Earth*;
> But claim superior Lineage by my SIRE,
> Who warm'd th' unthinking Clod with heav'nly Fire:
> Essence divine, with lifeless Clay allay'd,
> By double Nature, double Instinct sway'd.[7]

This depiction of man as son of earth and heaven communicates Arbuthnot's belief that man is more than a corporeal entity: he is a unity of the material ('unthinking Clod') with heavenly spirit, a divine animation of 'lifeless Clay'. The 'Sparks of heav'nly Fire' which animate man represent an attempt to avoid a reduction of mankind to mere material, mechanical matter:

'Essence divine', a divine animation, names precisely that aspect of man's nature which is both immaterial and gives life. Such an image, and a similar vocabulary of animation, recur throughout the texts examined in this chapter and beyond, but the role and significance of the divine spark evolves, until the animating principle becomes less a theological concept than a physiological one. In Book IV of John Armstrong's *The Art of Preserving Health*, published a decade after Arbuthnot's poem, this shift is already anticipated. The spark of 'immortal fire' which 'animates and moulds' the human frame is already more than an animating principle, operating as a ruling principle of health, sensitivity and feeling:

> this heavenly particle pervades
> The mortal elements, in every nerve
> It thrills with pleasure, or grows mad with pain.
> And, in its secret conclave, as it feels
> The body's woes and joys, this ruling power
> Weilds [sic] at its will the dull material world,
> And is the body's health and malady.[8]

Armstrong's principle of health anticipates later physiological attempts, central to vitalism, to theorise a nervous principle of animation as both a part of and a supplement to the material mechanisms of the body. The theological commonplace of animated clay, which is both the point of resolution and the unstable faultline in Arbuthnot's poem, is developed in Armstrong's text to begin to envisage a wholly different conceptualisation of man's nature. The account of such developments belongs more properly to later chapters, but the beginnings of what would later be more formally addressed in vitalist theories can already be seen in the fascinations and difficulties of attempts to understand man's animated nature in the earlier eighteenth-century writing examined in the rest of this chapter.

Natural philosophy and morality in the Scottish science of man

The eighteenth century's interest in the human subject received one of its most focused expressions in the 'science of man' which grew out of Scottish moral philosophy, both within and beyond the universities, from the early eighteenth century, and which was a key manifestation of the intellectual energies and ambitions of the Scottish Enlightenment. The 'science of man' developed from traditional moral philosophy by emulating the methodology of empirical natural philosophy, and a discipline which had largely focused on the religiously founded teaching of practical ethics was transformed to encompass analysis of the nature of man and his activities in a range of different fields. Francis Hutcheson's *A Short Introduction to Moral*

Philosophy (1747), a work probably closely based on his moral philosophy lectures at Glasgow University, thus begins with 'Ethicks', exhorting the traditional virtue of regulating the passions, but extends into the discussion of private rights, or natural liberty, 'œconomicks', or the laws and rights of members of a family, and 'politicks', an examination of various plans of civil government and the relationships between states.[9] Such an ordering of knowledge, which places the nature of man as the centre and foundation of what would now be seen as the human sciences, and addresses mankind's political and social actions from a foundational account of human nature, is repeated in Hume's introduction to his *Treatise of Human Nature* (1739), where the 'science of man' clearly places man as the common object of all branches of knowledge. With knowledge of mankind as their shared goal, different disciplines or forms of enquiry clearly overlapped – and the influence of the science of man was felt beyond philosophy too, including in literary culture. Its effects are evident in the fascination with human nature evident in all the writers discussed in this chapter, who share various links to Enlightenment Scotland. Arbuthnot and John Armstrong, although doctors practising in London, were both educated at Scottish universities where the new learning, as Hutcheson demonstrates, quickly found its way into the curriculum. Indeed, Arbuthnot's early connections to Newtonianism in early Enlightenment Scotland, an important influence on the new moral philosophy, complicates his later reputation as the satirical scourge of new science.[10] Thomas Blacklock, meanwhile, although born to English parents, grew up in Dumfriesshire and was educated at Edinburgh University, and the discussion of his plight by Hume and the novelist Henry Mackenzie exemplifies the Scottish philosophical concern with man's sympathetic impulses towards others, a central component of Scottish moral philosophy's theorisation of human sociability. It was also in this context, of burgeoning enquiry into human nature across a range of disciplines, that vitalist physiology emerged, to offer a new language by which man's physical embodiment, so central to questions of human subjectivity, might be understood.

Recent work has done much to uncover the influence of natural philosophy, and especially Newtonianism, on Scottish moral philosophy in the early eighteenth century. In broad terms, a key consequence of this influence was the adoption of 'scientific' methods of enquiry analogous to those of the new science – for instance, in constructing 'man' as an object to be observed and analysed in the objective and empirical fashion of natural philosophy. As William Leechman, his colleague at Glasgow University, explained, Hutcheson, a leading figure in the new 'science of man', hoped to enquire into 'the various natural principles or natural dispositions of mankind, in the same way that we enquire into the structure of an animal body, of a plant, or of the solar system'.[11] The adoption or emulation of a scientific method of analysis and synthesis, of division into parts in order to understand the whole, when applied to the human subject, typically constructed 'man' as an observable unity made up of interconnecting parts, independently operating, but together contributing to some collective end. If, for Leechman, there is 'one proper method ... to be followed in the moral science', it is 'to inquire into our internal structure as a constitution or system composed of various parts, to observe the office and end of each part, with the natural subordination of those parts to one another, and from thence to conclude what is the design of the whole, and what is the course of action for which it appears to be intended by its great Author'.[12] Just like the natural world studied by natural philosophy, 'man' as studied by the new moral science was a perfect mechanism whose design expressed the intention of its Creator; to understand that design was to enable the mechanism to operate in accordance with God's will.

But understanding man's moral nature as a kind of mechanism paradoxically brought a new emphasis on reason and the will, as in their absence a materialist account of man as a machine distinctly threatened. Mechanistic metaphors transported from natural philosophy into moral writing could thus unexpectedly work to highlight the question of the life-force, or animating presence, controlling and regulating the system. The question of who or what was being animated, and for what end, thus constantly recurred. When David Fordyce, Professor of Moral Philosophy at Marischal College, Aberdeen, depicted the passions through a mechanical vocabulary of springs, movements and motions, this in turn prompted attention to the 'Directing' and 'restraining' powers of the conscience, and Fordyce is drawn into theorising not simply a system of force, impulse and motion, but of moral control and animation:

> the passions are mere Force or Power, blind Impulses, acting violently and without Choice, and ultimately tending each to their respective Objects, without regard to the Interest of others, or of the whole System. Wheras the Directing and Judging Powers ... are capable of directing or restraining the blind Impulses of Passion in a due Consistency one with the other, and a regular Subordination to the Whole System.[13]

Fordyce's account of man's moral nature posits a system of blind mechanical forces focused on their own ends, overseen by a power of direction and judgement. But even with its emphasis on the power of conscience, a 'mechanistic' account of virtue, as analogous to the smooth running of machinery, never seems far distant, especially when comparisons are made between the purpose of man's moral system and the 'ends' of other objects of art. Like them, Fordyce asserts that man must be understood according to his intended design:

> Could we understand a Watchmaker, a Painter, or a Statuary ... should they tell us, that a Watch, a Picture, or Statue, were good when they were

true ...? Would they not speak more intelligibly, and more to the Purpose, if they should explain to us their End or use, and in order to that, shew us their Parts both together and separately, the Bearings and Proportions of those Parts, and their Reference to that End? Will the Truth, the abstract Natures and Reasons, the eternal Relations and Fitnesses of Things, form such Detail?[14]

Fordyce had begun his account of man's moral nature with an insistence that he would describe man not as he should be, but as he is, but such rhetoric of 'ends' reintroduces the exhortations to cultivate virtue associated with a more traditional moral philosophy teaching, and returns an account of man's 'system' to an older concern with practical ethics.

Hutcheson, the leading moral philosopher of his day, had defined the pursuit of 'happiness' as the discipline's main object, and, as with Fordyce, this turns out to be the efficient running of a perfect mechanism which both already exists and should be cultivated. We should, according to Leechman's description of Hutcheson's teaching,

> cultivate that temper of mind, and pursue that course of life, which is most correspondent to the evident ends and purposes of his divine work-manship; and that such a state of heart and plan of life, as answers most effectually the end and design of all the parts of it, must be its most per-fect manner of operation, and must constitute the duty, the happiness, and perfection of the order of beings to whom it belongs.[15]

Man is both already a piece of perfectly designed 'workmanship', and must regulate himself to become one: a conclusion which at once acknowledges an act of divine creation, and exhorts mankind to both complete and supplement God's workmanship with their own acts of self-production. A language of machinery, system and design in part derived from natural philosophy is thus reconciled with the traditional exhortations to virtue of an older moral philo-sophical tradition. But this sense of man's nature as a system whose perfect operation might only be ensured by an overseeing regulating force betrays the weakness of a mechanical model as an analogy for man's moral nature: such machinery cannot be allowed to operate without a regulating guiding hand, or some kind of vital presence. At the same time, the desire to understand man as in some way an organised system is strongly evident. Vitalism's metaphors, which retained the sense of an organised system, but dispensed with the prob-lematic tension between blind forces and a controlling function, would offer a fruitful alternative formulation which, as the next chapter argues, would also inform the formulation of sympathy in Scottish moral philosophy.

An alternative metaphor frequently deployed in Scottish moral philosophy also demonstrates the influence of an analogy with natural philosophy: the image of the science of man as an 'anatomising'. Analysing man's nature was repeatedly described as akin to a physical dissection: as a body cut, opened and exposed to trace its structure, organisation and secret operations. The prevalence of such anatomical images again helps to explain how vitalism's new physiological language could so readily be absorbed into moral philosophy's accounts of the human subject. Fordyce's *'particular* View' of man is offered as a dissection, an *'Anatomy'* of an 'inward and more elaborate Subject' in which, although 'it will not be necessary to pursue every little Fibre, nor to mark the nicer Complications and various Branching of the more minute Parts', the 'larger Vessels and stronger Muscling of this Divine Piece of Workmanship' will nevertheless be laid open.[16] Such images point to the complex relationship between a science of man and other forms of natural enquiry, with moral philosophy's object, the nature of 'man', being reified under the influence of its analogy with the science of nature. Recurring formulations, like Hutcheson's determination to 'search accurately into the constitution of our nature, to see what sort of creatures we are', or to 'expect to find in our structure and frame some clear evidences, shewing the proper business of mankind', exploit a metaphorics of the body to suggest a visible, and so investigable, embodiment of their subject.[17] Such images at times lie alongside more mechanical depictions of man as a piece of 'Machinery' – Fordyce's proposed 'Anatomy' comes within a few lines of a description of the *'Cogs'* and 'delicate *Springs'* by which man is *'impelled* to Action' – as though to underline the difficulty of picturing the science of man's object, or as though an image of the body as machine underlies and reconciles the apparently mixed metaphors. Elsewhere, Hume's determination that a moral science is precisely a kind of, as the Abstract of his *Treatise* puts it, anatomising of 'human nature in a regular manner', expresses the difference of his own conception of the science of man as an unbounded scientific enquiry from Hutcheson's insistence on painting the beauties of nature:

> There are different ways of examining the Mind as well as the Body. One may consider it either as an Anatomist or as a Painter; either to discover its most secret Springs & Principles or to describe the Grace & Beauty of its Actions. I imagine it impossible to conjoin these two Views. Where you pull off the Skin, & display all the minute Parts, there appears something trivial, even in the noblest Attitudes & most vigorous Actions: Nor can you ever render the Object graceful or engaging but by cloathing the Parts again with Skin & Flesh, & presenting only their bare Outside. An Anatomist, however, can give very good Advice to a Painter or Statuary: And in like manner, I am perswaded, that a Metaphysician may be very helpful to a Moralist.[18]

Hume's evocation of the anatomist's repeated peeling away of layers and cutting through of surfaces depicts man's nature not as an object of beautiful contemplation but as something painfully but necessarily exposed.

Unlike a Hutchesonian enquiry, concerned to discover man's purpose or 'End', moral science is here an enquiry bounded only by the extent of man's nature itself, by the limitations of possible division and dissection, an enterprise whose potentially reductive nature was to haunt Hume's personal accounts of his philosophical labours. The parallel with anatomical science here constructs the body as the knowable object of a heroic science, but, as we have already seen in the Introduction, elsewhere in Hume anatomy figures in a far more sceptical way – as a final stage of possible knowledge before we arrive at a sense of the 'mysterious and unintelligible' power, the 'inward sentiment or consciousness', which animates the body.[19] Here anatomy stops short of revealing the real hidden powers of the body – whose nature vitalism will soon postulate. Anatomy as a model for the science of man can ultimately only offer partial knowledge, which vitalism, in its description of possible animating forces, hoped to complete.

The use made by eighteenth-century Scottish moral philosophy of a range of bodily images in the organisation and representation of its own pursuit of knowledge, demonstrates the extent to which it was inspired and indebted to contemporary natural philosophy, where enquiry into empirical objects, including human physiology and anatomy, was a central activity. This fascination with the images and metaphors suggested by the body, as well as the fruitful figurative language and analytical concepts associated with it, suggests too how readily a new bodily discourse, of animation rather than mechanism, might be picked up and assimilated by moral philosophy on its emergence in the 1740s. The development of the concept of sympathy, which in early eighteenth-century moral philosophy is relatively unexplored, into a more powerful animating force in Adam Smith's *Theory of Moral Sentiments*, as explored in Chapter 2, is one sign of this. The creativity implicit in registering and redeploying such concepts and terminology in a new mode, meanwhile, also helps to explain the adoption of bodily language by other, more literary forms of writing. But if the material metaphors which often resulted from looking to natural philosophy offered the attractions of a 'scientific'-seeming analysis, they potentially also produced a godless account of man, as his nature is reduced to the mere operation of a system. This is the debate which was played out in an important exchange between Hutcheson and Hume, which exposed the tension between the desire to retain a sense of man's divine 'beauty', and the opposing impulse to submit it to an unremitting intellectual dissection. Such tensions are not fully dissipated with the emergence of vitalist physiological theories at mid-century, but are recast in a new debate, between those for whom man's 'vital principle' can be assimilated to an animating 'divine spark', and those for whom such language merely repeats a tendency to materialism already present in a moral philosophy too closely allied to the new science. Although it was not until the end of the century that the politics of vitalism were fully articulated, the divisions which were then exposed continued a debate which began with alternative interpretations of the purpose and methods of a newly analytical moral philosophy in the Scottish science of man in the first decades of the century.

John Armstrong's bodily subjects

In the philosophical context of the Scottish science of man, the body operates primarily as a metaphorical resource, a means of offering material images of man's moral nature. *The Art of Preserving Health* (1744) by John Armstrong (1708/9–79), a Scottish physician and poet who pursued medical and literary careers in London (at one stage attending Fanny Burney), illustrates a comparable curiosity about the human subject in the very different context of the health-advice poem. Armstrong's poem was widely known, going through 11 London editions between 1744 and 1796, as well as others in Dublin and North America. Armstrong provides another way of 'knowing yourself', and becoming yourself, from that explored by Arbuthnot and the moral philosophers: not via religious faith or philosophical enquiry, but through proper 'lifestyle' regimens of diet, exercise and sexual conduct. His refusal of 'higher' theological or philosophical accounts of the self and modes of self-knowledge is all the more evident in the self-reductive parallels with Miltonic epic explored by *Preserving Health*: Armstrong's poem is not a parodic but a reduced epic which reveals the ways not of God, but of health, to man, transferring the grandeur of epic celebration, classical reference and other epic machinery to the exposition of a modern, secular knowledge of man by man.

Yet although the poem presents such mundane activities as digestion as an effort of epic magnitude – 'a work / Of strong and subtle toil, and great event: / A work of time' (III.352–4) – or animates a discussion of hangovers with mythological references (IV.190–3), its account of the body and health is informed by, and attempts to intervene in, prevalent contemporary debates and anxieties about luxury, excess, indolence and dissipation. In this context, Armstrong uses the body both to demonstrate the threat posed to a heroic masculinity by the pervasive influence of an effeminising, indulgent modernity, and to formulate a countering model of order and regulation, of proper exercise and labour in the face of such insidious dangers. In some ways then, even when the body appears to be at the forefront of Armstrong's concerns, it is the least of them, operating as a means through which larger social and cultural anxieties can be articulated and exorcised, and constructing a subject as both receptive to and resistant of cultural 'contagion'. Despite his evident medical expertise as a physician, Armstrong's body thus operates in a fruitful realm somewhere between knowledge, speculation and fantasy, where current anatomical and physiological discoveries and hypotheses give rise to suggestive imaginings in which the bodily – its regulation, exercise, governance and organisation – defines a key point of interchange between subject and culture, constructing a subject whose physical being defines the terms of his interaction with, and participation in, the currents and forces of the larger social world.

Most importantly for our concern with narratives of embodiment, however, Armstrong offers an often unresolved depiction of the mind's relation to the body, sketching different versions of this at different points in the poem. A conventional emphasis on the mind's overruling control, such as Fordyce had insisted on in his moral philosophy, competes with periodic acknowledgements of bodily independence, an account which anticipates the later emancipation of the body from mental rule in vitalist physiology. The oscillation between these two alternatives marks a fundamental uncertainty in the text about the nature of, and relationship between, mind and body, as though the poem represents a shift (or the beginning of such a possibility) from an earlier account of the body as ruled by and the instrument of the mind, and later inclinations to view it as an independent and self-directing system. As we have already seen, Armstrong's attention to the health-securing properties of the blood in itself looks beyond a dualist model of a body overseen by the mind, towards a 'heavenly particle' pervading every nerve as a 'ruling power' co-extensive with each 'secret conclave' of the body. The blood would also be considered the location of the mysterious vital principle by late eighteenth-century anatomist John Hunter.

Armstrong's prescriptions for the 'preservation' of health, which is thus from the outset constituted as under threat, define health as a happy medium between opposed poles of vigour and indolence, discipline and indulgence. His recommended regimen of properly balanced exertion and rest, consumption and restraint, can be understood conventionally enough within a well-documented context of eighteenth-century fears about luxurious modernity, where traditional political virtues and values appeared jeopardised by the emergence of a credit-based economy which seemingly licensed desire, encouraged excessive consumption, legitimised vice and undermined established social hierarchies.[20] Armstrong's poem is rendered anxious by a profusion of choice, wealth, luxury and excess, depicting modern man as caught in an age where the plenitude of nature results in chaotic and unordered consumption; where the robust health consequent on physical labour is available only to the working classes; where the energising activity of conquest is confined to history; and where rich diet undermines life and longevity. The dissipation of self threatened by all such cases is figured as a danger specifically to the physical self, and as in *The Castle of Indolence* (1748) by Armstrong's friend James Thomson, to which Armstrong himself contributed descriptions of Lethargy, Hydropsy and Hypochondria, physical lassitude signals larger spiritual, moral and economic failings. But in *Preserving Health*, the body can also at times provide a remedy to such extra-corporeal concerns: its forceful, potentially destructive, operational energies invigorate a subject otherwise lost in passivity and lassitude, as the poem unleashes an array of advice and prescription to transform indolent hours to active ones. In an anticipation of the vital energy later foregrounded by Whytt and others, the innate, vigorous, vital power of the body is thus a necessary counter to an opposing, culturally induced tendency to inertia. The poem's attention to the physical efforts of such bodily operations as digestion, blood circulation, breathing and perspiring, thus foregrounds the often unconscious vital activities and energies of the body as safeguarding a health often perceived as under threat from wider social and cultural contexts.

The cultural and moral fears which are represented and countered in the poem are presented in Armstrong's account of nature, so that the self-definition of the body against the insidious corruptions of nature becomes a central concern. In its focus on the 'non-naturals' of health – air, exercise, diet – *Preserving Health* pays special attention to the relationship between the internal 'economy' of the body and its external context, a relationship which, because undefined and unregulated, is particularly problematic in Book I of the poem. Here, the health of the body is threatened by poisonous and diseased elements of both the natural world and the man-made environment, its vulnerability emphasised by the body's necessary internalisation of elements of the external. Breathing the 'turbid air' and 'volatile corruption' in the 'rank city', for instance, is hazardous, and is only rendered safe by the unintendedly benign action of chimney smoke, which, in Armstrong's confused physics, dries out the otherwise dangerously moist air. In a context where health is understood as a balance between extremes of wetness and dryness, the danger is that the body internalises the air's excessive moisture, a danger heightened by the presentation of its organs as passive and dependent, 'drawing', 'yawning', 'drunken' and 'drinking'. Such terms personify, and thus animate, the body's organs and parts – its cells, 'venous tubes', lungs – yet the extension of human qualities to physical mechanisms only underlines their inability to exert conscious control over the body as a whole: it is not through their efforts, but the chance drying of air, that the heart, in another personification, is prevented from being 'roused' to 'rage'. But whilst personification of the parts suggests a certain level of animation and independence, the efforts of these body parts only emphasise the passivity and disunity of the body as a whole, lacking as it does both a mental overseeing regulatory power, and a unifying bodily principle of automatic self-preservation.

Armstrong's conception of health as a happy medium between dangerous extremes means that it is not only moisture and humidity which pose a threat. An excessively dry climate can also infect a body all too vulnerable to a passive repetition of the qualities of its surroundings, so that the internal workings of the body form an interior landscape which microcosmically repeats the qualities of its exterior surroundings:

> Spoil'd of its limpid vehicle, the blood
> A mass of lees remains, a drossy tide

That slow as Lethe wanders through the veins,
Unactive in the services of life,
Unfit to lead its pitchy current thro'
The secret mazy channels of the brain. (I.173–8)

In this pathology, disease is the failure to actively manage one's relationship with the environment, of allowing the body too passively to imbibe and repeat the characteristics of its surroundings. The labour exhorted in this health-georgic is that of a necessary self-governance, a controlling and managing of the body's relationship with what is exterior to it in a way which enables the body to transform the exterior – air, food, environment – into its own health and life. Armstrong's construction of the body as prone to the passive absorption of outside atmospheres or environments thus begins to point to the need for an overruling, controlling function. But at the same time, a sense emerges of the self-regulating capacities of the body, which render such overruling functions of control potentially redundant, and eventually it is the vital force of the blood which proves the final guarantor of bodily health.

This increased sense of bodily independence is especially evident in the elaboration of an essentially economic relationship between the body and external world, which develops from Book II. Such a relationship operates as a means of ordering and controlling the plenitude of nature which in Book I is experienced in passive and troubling ways. Alongside its strictures about restraint, balance and moderation, the poem elaborates an account of the body as an economy: a system of exploiting and transforming the resources of nature which are then circulated productively around the body. It is an image which suggests order, process and purpose in an otherwise excessive and meaningless profusion of nature, and which, as well as defining and controlling the body's relationship with the exterior world, also depicts the body itself as efficient, active and productive. All this is evident in the representation of digestion as an organised system of production and distribution. Digestion is presented as the production of 'chyle' or digestive juice, which is carried around the body in blood:

the concoctive powers, with various art,
Subdue the cruder aliments to chyle;
The chyle to blood; the foamy purple tide
To liquors, which thro' finer arteries
To different parts their winding course pursue;
To try new changes, and new forms put on,
Or for the public, or some private use. (II.35–41)

This vigorous, beneficial circulation of digestive juice becomes a recurring image of health and indeed life itself: blood is described as a 'vital fluid', its circulation a 'vital force' (II.16, 27). Such a systematisation of the body into a productive economy offers not just a regularised model of the process of consumption, but an account of unconscious, independent, vital functions whose actions ensure the health and vitality so necessary to life.

At its most ideal, the economy formed between the body and external nature makes a perfect system: a benevolent, beneficent correspondence of needs and wants, a microcosmic manifestation of the divine harmony and order of the universe as a whole. At such moments, the body's natural system does appear to be – like those in vitalist physiology – an independently self-operating entity. At the same time as already existing, however, this perfect mechanism, like the mechanisms of man's moral nature elaborated by Hutcheson and Fordyce, must be cultivated and supported through precisely such regimens of diet, exercise and so forth that Armstrong elaborates. This tension, between a body operating independently under 'nature's wise oeconomy' (II.133), and one needing conscious governance and regulation, is focused through the presentation of waste. The poem recognises some 'unavoidable expence of life', such as the 'necessary waste of flesh and blood' expended in repairing the body's physical fabric, damaged by the 'hourly' battering of the circulating blood (II.33–4), or the 'expence' of excremental surplus, maintained by nature '[w]ithout or waste or avarice' (II.133–4). Other forms of waste are less easily incorporated into a model of an independently self-contained system, and argue the need for some form of external regulation to control what appears a dangerous loss or dissipation of self. Perspiration, for instance, presents a picture of inevitable and ceaseless dispersion of fluids through the 'endless millions' of 'small arterial mouths' which 'pierce' the skin, a microscopic image shocking in its sudden focus on minute but magnified and uncontrollable dispersion (III.256–9). At its 'wonted measure', perspiration moves the 'wheels of life' (III.262–3), but the passage is haunted by a fear of self-dissipation, heightened by Armstrong's understanding of the process as the loss of health-giving blood. Intervention, supervision and control are again necessary as Armstrong interdicts the relaxing baths and oils in which the luxurious Romans had indulged; in our own 'less voluptuous clime', the cultivation of too 'soft' a skin might threaten the very coherence of the self, just as, later on, sex is feared as a depletion, even dissipation, of the self. Whatever the attractions of elaborating a wholly independent, self-directing bodily system, residual fears about bodily collapse reintroduce the themes of supervisory regulation and good governance: nature's independent operation can finally, for Armstrong, only take place under the overseeing eye of human control.

Ultimately, then, Armstrong resists the possibility of independent bodily self-regulation and turns instead to a political model of governance, finding in the detail of a somatic economy a lesson about control and regulation, just as Hutcheson and other moral philosophers offered exhortations to self-cultivated virtue in their accounts of man. The activities on which he focuses – breathing, digestion, perspiration – underline the body's potential vulnerability, poised in a necessarily dependent relation to the external world, to which a defensive attitude is advised. Although the internal workings of the body occasionally call for a form of suppressive governance – blood is often enraged, needing to be subdued – the poem's energies are largely directed at the potential threats posed by the exterior. Armstrong's exhortation to cultivate a (literally) tough skin – necessary, he says, if you are to participate in trade or indeed to socialise, without catching disease – exemplifies the poem's instinctive modelling of what is exterior to the body as a potential threat, and its inability to imagine less defensive forms of exchange. In contemporary Scottish moral philosophy, the word 'contagion' often describes a sympathetic and automatic sharing of emotion between persons, an inevitable sociability and exchange, but for Armstrong the same word, whilst still describing man's relationship to man, remains unalleviated by any benevolent sympathies, and suggests a potentially injurious relationship, a pathology of human contact. The self-sufficiency and defensive self-isolation implied by this is underlined by the consideration that Armstrong's bodily economy works to produce what it already has, namely health and life; the labour which Armstrong celebrates and enjoins in this georgic poem thus appears strangely circular, its productive energies less generative of the new than sustaining of what is already in existence. If the exhortations of Hutcheson and others to cultivate virtue holds out a model of cultivation as self-improvement, Armstrong enjoins a redundant, unnecessary labour, which is in fact already taking place, to produce what already exists. But not to insist on the necessity of that labour would mean suggesting a body best able to look after itself independently.

Armstrong's narrative of the body thus insists on a regulating, labouring self. But this relationship between subject and body is disturbed when Armstrong's understanding of the body as a material organisation overseen by the will reaches its limits. In Book IV, the poem's informing model of a body regulated and overseen by human intervention is modified to suggest a divine 'spark' which 'animates and moulds the grosser frame', a 'heavenly particle' or 'ruling power' which 'weilds [*sic*] at its will the dull material world, / And is the body's health and malady' (IV.12–13, 16, 20–2). The image recalls Arbuthnot's 'lifeless clay', animated by 'heav'nly Fire', but also suggests a mysterious health-giving principle, a 'subtle principle within' which '[i]nspires with health, or mines with strange decay / The passive body' (IV.5–7). Such a notion, combining divine animation with health-giving powers, troubles the model of a humanly administered body which has sustained the poem thus far, rendering potentially redundant the 'choices', of aliment, air and toil, evoked in the book's opening lines. The poem's mutually supporting relationship between body, health and self is thrown into doubt, displaced by the possibility that health is prescribed by some combination of natural and divine cause. The human subject, hitherto understood as the necessary regulator of the physical body which it inhabits, is made strangely absent, as Armstrong sketches the body's materiality under some form of divine or natural control. As with Arbuthnot, we meet again an unresolved tension between the body as self-determining and as divinely controlled. The themes of displacement and dislocation, which had so haunted Arbuthnot's attempt to 'know himself', recur here, not with a human self ejected from its 'native seat' in heaven, but through a final inability to define the relationship between an animated body, the material causes of its health or illness, and the self who inhabits it.

In a poem whose sustaining premise is the offering of advice, and whose greatest fear is the collapse of health through inadequate somatic management, the sketching of an overruling divine or natural principle of health precipitates a crisis in the very idea of regulation. It opens a vision of a body operating naturally and healthily whilst independent from human direction – a vision which distances man from nature and emancipates the physical world from human intervention and control. Such possibilities strikingly anticipate the vitalist physiology which saw health and life as ultimately regulated by a mysterious and inaccessible 'life principle' – a principle which rendered such lifestyle prescriptions as Armstrong's unnecessary. The underlying emphasis on governance, control and regulation which sustains Armstrong's bodily narrative in this poem, however periodically threatened, together with the poem's construction and naturalisation of human authority over the administered 'nature' of the body, suggests just how politically suggestive (and for some, politically troubling) such an alternative formulation of a bodily economy might be. In Armstrong's poem such possibilities are just held off, but they were to reappear later in the century, as we will see in subsequent chapters.

Thomas Blacklock's blindness: the body impaired

The three contexts examined thus far demonstrate something of the variety of ways in which man's material embodiment is written in eighteenth-century Scotland, whether in Arbuthnot's articulation of man's mixed heavenly and fallen nature via a mind–body split, in Scottish moral philosophy's mapping of man's moral nature through mechanical bodily metaphors, or in Armstrong's oscillation between alternative modes of bodily regulation or independence. The case of Thomas Blacklock (1721–91), a humbly born, Edinburgh-educated poet, blind from the age of six months, who was celebrated by David Hume and Henry Mackenzie, offers a further twist to eighteenth-century accounts of man's corporeal nature, which both continues and confuses assumptions about the ways in which the body informs a sense of self. In Blacklock's case, however, it is an imperfect body which prompts a reconsideration of the relationship between subjectivity and the corporeal. Writings on Blacklock by contemporary commentators are preoccupied by the ways in which his blindness informs and constructs his literary productions, his social relations, and even his temperament, manners and character, so that his entire person comes to be defined by his impaired body. Although partly constrained by what one writer refers to as the 'impropriety' of referring to yourself in public discourse, Blacklock's own writings also suggest the ways in which his sense of identity is profoundly informed by the particular nature of his embodied state. Blacklock vigorously contests a self defined and, as he perceived, imprisoned by bodily impairment, and attempts in his poetry to defy the constraints of his physical state, both through the production of the visually descriptive verses which his blindness would appear to preclude, and through his repeated invocation of figures and acts of both poetic and religious transcendence. But where, in Armstrong or Arbuthnot, the experience of the limitations of the material mechanisms of the body can at times prompt a search for a different kind of language to speak of some form of vital or animated being, such a move is not made in Blacklock's own writings about his impaired physical being, which turns instead to the register of melancholy and lament. For writers in this chapter, at least, a discourse of vital animation is not an automatic alternative to narratives of the proper functioning of bodily mechanism, and even sensibility, offered by David Hume as a form of compensation for Blacklock's blindness, is not the more fully fledged vital force which, as the next chapter shows, it can be elsewhere.[21]

The relationship between sensibility – or some versions of it – and vitalist physiology are explored in the next chapter, where we see sensibility, although never quite equivalent to it, is often discussed in ways which explore its similarity to a kind of vital organic force. Sensibility appears rather differently in the discussions of Blacklock by David Hume and other commentators, where Blacklock's sensibility is a mark of his social difference – at times, almost grotesquely so. As something which 'compensates' for the disability of his blindness, Blacklock's sensibility looks back to the failed mechanisms of his bodily senses, and, far from a transformative vital energy, could almost be understood as his physical disability manifested again in another form. If sensibility in such a context never really approaches the sense of the powerful force it can be elsewhere, it nevertheless marks an attempt to supplement the mechanisms of bodily senses (Blacklock's failed sense of sight) with some kind of non-mechanical, non-bodily force – thus repeating the same need to look beyond basic bodily mechanisms as we have already seen in Arbuthnot and Armstrong. Such narratives of a compensatory sensibility, strongest in Hume's reading of Blacklock, are countered, however, in Blacklock's own accounts of blindness, where the peculiar sensitivity consequent on the loss of sight is not an exemplary social feeling, but a fearful and heightened sense of vulnerability and dependence. Where Hume compensates Blacklock's loss of sight with the attribution of extraordinary powers of artistic sense and social sensibility, Blacklock's mournful account of blindness suggests a melancholic but impossible desire for the return of sight itself – resisting even the limited forms of alternative powers which Hume's version of sensibility offers.

As a blind man who produced poetry which often included extended passages of visual description, Blacklock attracted much amazed comment following the publication of his first collection of verse in 1746; some commentators were moved to offer explanations of how such a seemingly impossible phenomenon had taken place. At a time when empirical models of human understanding stressed the role of the body's senses in acts of perception, the 'sighted' blindness apparently evident in Blacklock's poetry disturbed conventional models of human apprehension. Blacklock's poetic descriptions offered the paradox of apparent visual perception without sight; and even when these were later explained through Blacklock's particularly receptive reading and memorising of other poets, his poetry still represented a visual description without an originary act of sight, and the assertion, against received empirical wisdom, of the independence of memory and imagination from the senses.

Commentators were quick to offer explanatory narratives to account for Blacklock's poetry. In his introduction to a posthumous issuing of Blacklock's collected works, Scottish novelist Henry Mackenzie, for instance, elaborated a theory of vision which did not depend on sight. Blacklock's case certainly represented a philosophical quandary, and reopened questions of the relationship between the senses, the mind and language, the subject of a philosophical debate to which Locke, Berkeley and Burke, among others, had contributed. The most elaborate explanation of Blacklock's poetic 'sight' was offered by Joseph Spence, Professor of Poetry at Oxford, who wrote a preface to a collection of Blacklock's work issued in 1754 and 1756. Here, Spence speculated that the poet, like other blind persons, may have developed the ability to distinguish colour by touch: '[t]he very same variety in the disposition of the parts in the surfaces of objects, which makes them reflect different rays of light to the eye, may make them feel as differently to the exquisite touch of a blind man'. By naming these 'different sorts of feelings', such a person 'may make a new sort of vocabulary to himself' and thus name the same colours as the sighted, even whilst his sensations of those colours would be very different.[22] Spence's narrative speculatively weaves the fragments of natural philosophy known by any reader of *The Spectator* – that colour is a consequence of the physical nature of an object, rather than inherent in it – into a fanciful extension of the blind man's touch, deemed so 'exquisite' as to be able to feel an object's capacity for light-absorption. It compensates for blindness by postulating the possibility of sight through another bodily sense – retaining an empirical model even whilst necessarily developing the blind man's touch to an 'exquisite' level beyond his fellow creatures. Spence's investment in this explanation, with its substituted account of the mechanism of touch for that of sight, was such that it remained in his preface even in its second edition, issued after a meeting with Blacklock at which the poet refuted Spence's speculations entirely, as Spence's footnote does at least concede. Even the personal testimony of the poet himself was clearly not sufficiently authoritative to counter such a compelling narrative of the peculiar talents of the blind.

In attempting to ascribe to Blacklock some form of supplementary ability, Spence was in line with Enlightenment thinking on blindness, which, as Katie Trumpener has noted, often understood the mind as attempting to compensate for its deficiency in sight through particular ability in other faculties.[23] Spence's suggestion that the blind form an associative link between touch and language also recalls Locke's anecdote, in his *Essay Concerning Human Understanding*, of a blind man who associates the word for the colour scarlet with the sound of a trumpet. Blacklock himself readily admitted to forming similar associations between colours and 'moral' concepts, so that the word 'green' suggested 'ideas of peace and serenity' and 'red' suggested courage.[24] Such forms of associative thought were considered at length in Hume's *Treatise of Human Nature*, but Hume's own account of Blacklock turns to a different register. Cast as an exemplary and moving instance of 'virtue in distress', a phrase actually used by Hume to describe Blacklock, the poet is lauded as the embodiment of social values and virtues whose currency social commentators in the 'age of sensibility' sought to secure. But Blacklock's sensibility, whilst in one sense exemplary, turns out to be more disabling than empowering.

Hume's account to Spence, in 1754, of his first meeting with Blacklock, offers a detailed construction of Blacklock's double-edged sensibility:

> The first time I had ever seen or heard of Mr. Blacklock was about twelve years ago, when I met him in a visit to two young ladies. They informed me of his case, as far as they could in a conversation carried on in his presence. I soon found him to possess a very delicate taste, along with a passionate love of learning. Dr Stevenson had, at that time, taken him under his protection; and he was perfecting himself in the Latin tongue. I repeated to him Mr. Pope's Elegy to the Memory of an unfortunate Lady, which I happened to have by heart: and though I be a very bad reciter, I saw it affected him extremely. His eyes, indeed, the great index of the mind, could express no passion: but his whole body was thrown into agitation. That poem was equally qualified to touch the delicacy of his taste, and the tenderness of his feelings. I left the town a few days after; and being long absent from Scotland, I neither saw nor heard of him for several years. At last an acquaintance of mine told me of him, and said that he would have waited on me, if his excessive modesty had not prevented him. He soon appeared what I have ever since found him, a very elegant genius, of a most affectionate grateful disposition, a modest backward temper, accompanied with that delicate pride, which so naturally attends virtue in distress.[25]

In this passage, Blacklock's 'bodily agitation' operates as an expression of the blind man's taste and sensitivity, prized socially acquired values which somehow appear both more extreme and inherent in him by their association with his body. Under Hume's benevolent experimentation, he demonstrates a kind of naturalised, corporealised taste in which sensitivity, to both the poem and Hume's condescension to him, manifests itself as an extreme physical response. In comparison, the conventional practices of politeness performed by Hume and the ladies in the passage, such as not discussing Blacklock in his presence, are overshadowed, even whilst the polite scene which is staged by Hume's visit, and repeated by the letter which reports it, itself enables and interprets Blacklock's dramatic display. But Blacklock's acute social ability – to respond with sensitivity to poetry – is also a kind of social disability: the uncontrollable bodily fits, which make him an extreme instance of the 'delicacy of taste' which Hume describes in his essay of that title, transgress expected norms of bodily comportment and, at least as the anecdote is reported, take the place of the polite conversation in which Blacklock might otherwise be expected to communicate his poetic enthusiasms. The 'excessive modesty' which prohibits him calling on Hume and thus benefiting from the forms of patronage which are eventually bestowed on him, is similarly both exemplary and disabling. The fact of Blacklock's blindness, his physical impairment, thus gives way in this passage to the consideration of the poet's *social* talents and weaknesses, where his exemplary qualities and virtues nevertheless maintain the sense of his disability which makes him a fitting object of Hume's concern.

Hume's reading of Blacklock, like other biographical accounts by Spence and Mackenzie, compensates the blind man by attributing to him the virtues of taste, character, sensitivity and politeness. Blacklock's sensibility, lauded by Hume, is thus a peculiar social ability, not the transcendent animating force or vital energy it could be for later theorists. But the limits of such 'compensatory' readings of his blindness are asserted by Blacklock himself, for whom, both in the few instances of its poetic treatment and in an article in *Encyclopaedia Britannica*, blindness is the subject of profound depression and melancholy. Writing in the third person, Blacklock's anonymous encyclopaedia article offers an apparently objective and authoritative celebration of the capacity of sight which suppresses the fact of its author's own blindness, and balances this with a lamentation of the situation of the blind, in which their visual deprivation carries heavy consequences for their entire physiological and psychological being:

> The sedentary life, to which by privation of sight they are destined, relaxes their frame, and subjects them to all the disagreeable sensations which arise from dejection of spirits. Hence the most feeble exertions create lassitude and uneasiness. Hence the native tone of the nervous system, which alone is compatible with health and pleasure, destroyed

by inactivity, exasperates and embitters every disagreeable impression. Natural evils, however, are always supportable; they not only arise from blind and undesigning causes, but are either mild in their attacks, or short in their duration: it is miseries which are inflicted by conscious and reflecting agents alone, that can deserve the name of evils.[26]

Blacklock's link between blindness and a dangerous lassitude, a necessary indolence and passivity, recalls Armstrong; yet where in Armstrong such inactivity must be countered by exercise and exertion, in Blacklock it must ultimately be accepted and submitted to, as a bearable 'natural evil'. The state of blindness is presented as one of peculiar passivity and vulnerability; the blind, as Blacklock asserts earlier, are 'rather to be considered as prisoners at large, than citizens of nature', condemned not merely to a state of sightlessness, but to larger, consequent forms of constraint and suffering which begin with that of the body and mind, and extend into other social, economic and cultural deprivations. It is especially the psychological vulnerability of the blind which is Blacklock's concern here, describing a sensitivity which calls for a special treatment by those around them. Such susceptibility, even sensibility, is particularly evident in the discussion of the care of blind children, and especially in his advice not to read them horror or ghost stories. For the blind, darkness and silence 'have something dreadful in them', and tales of ghosts, fiends or avenging furies 'seize and pre-occupy [sic] every avenue of terror which is open to the soul; nor are they easily dispossessed', especially as light and visual images cannot give them easy dispatch.[27] This depiction of the blind as caught in a horrific space of gothic terrors, unable, by their own physical incapacity, to free themselves from the furious creations of their imaginations, is the flipside of Hume's celebration of Blacklock's poetic susceptibility, or even of Spence's 'exquisite' touch: a peculiar, inevitable and uncontrollable sensitivity bequeathed by their blindness. But where, for other commentators, forms of sensibility in the blind are happy compensations for visual impairment, or iconic examples of sentimental culture's 'delicate taste', for Blacklock they only underline the essential dependence, physical and psychological vulnerability of sightlessness. The sensitivity of the blind, for Blacklock at least, does not open the door to a newly animating mode of being, but rather underlines the deficiency and lack consequent on the impairment already suffered.

The mix of fortitude and gothic terror, of submission and misery, which characterises Blacklock's encyclopaedia account of blindness recurs again in his two poetic treatments of the same topic, the 'Hymn to Fortitude' and 'A Soliloquy'. The first of these, a graveyard extravaganza which recounts the horrors of the 'fiends' of night emerging at midnight to strike horror into the breast, only to be despatched by resolution and 'fortitude', whilst not explicitly broaching the issue of blindness, nevertheless strikingly replays the same language and themes which will later describe blindness in the *Encyclopaedia*:

> Curst with unnumber'd groundless fears,
> How pale yon shiv'ring wretch appears!
> For him the day-light shines in vain,
> For him the fields no joys contain;
> Nature's whole charms to him are lost,
> No more the woods their music boast;
> No more the meads their vernal bloom,
> No more the gales their rich perfume:
> Still darkness thickens to his eye,
> Blots all the field, contracts the sky.[28]

'Blot' is a verb Blacklock uses elsewhere to refer to the cutting off of his sight, and sensory deprivation, horror, misery and isolation are all linked here, even if caused by fear rather than blindness. Especially when read, as the collection's prefatory material prompts us, as the production of a blind man, the passage suggests a gothic account of what the experience of blindness might be, associating sightlessness with fear, vulnerability and loneliness. Blindness is broached more explicitly in the 'Soliloquy', a poem written, as the full title communicates, on *'the Author's escape from falling into a deep well, where he must have been irrecoverably lost, if a favourite lap-dog had not, by the sound of its feet upon the board with which the well was covered, warned him of his danger'*. The comic and sentimental possibilities latent in this lengthy contextualisation, however, are resisted, giving place to a lengthy lamentation of the deprivations, horror and dangers of blindness:

> O Beauty, Harmony! Ye sister train
> Of Graces; you, who in th'admiring eye
> Of God your charms display'd, ere yet, transcrib'd
> On nature's form, your heav'nly features shone:
> Why are you snatch'd for ever from my sight,
> Whilst, in your stead, a boundless waste expanse
> Of undistinguish'd horror covers all?
> Wide o'er my prospect rueful darkness breathes
> Her inauspicious vapour; in whose shade,
> Fear, grief, and anguish, natives of her reign,
> In social sadness, gloomy vigils keep:
> With them I walk, with them still doom'd to share
> Eternal blackness, without hopes of dawn.[29]

Here there are no alleviating compensations to relieve a blindness experienced as '[f]ear, grief, and anguish', only an unremitting loss. Both the poems attempt resolution by appealing to fortitude and religious faith respectively, but although such moves suggest the poet transcending the sufferings which the poems describe, they in fact document his submission to them. In each case, by invoking strengths of faith and character, Blacklock highlights qualities regarded as exemplary by biographers and commentators, yet the effect is only to point out the limits of their 'compensatory' readings of his blindness, an observation which only underscores his own melancholic representation of blindness. Whilst fortitude and faith can counter misery and needless fear, good character cannot cure blindness.

Blacklock's poems identify the real affliction of blindness as a form of social exclusion, an insight which casts his disability not as a physical affliction but as an impairment of his ability to participate fully in eighteenth-century Scottish culture of sentimental sociability. For Blacklock, the significance of his body is not limited to the corporeal, but extends to the consequences his impaired body has for his participation in the social and cultural context to which his body belongs, and in which its afflictions are interpreted. By insisting that his real disability belongs to his social, rather than physical, being, Blacklock asserts the commonality of man's sociable nature as transcending mere facts of bodily difference; his physical difference only underlining that it is not physical identity, but the enactment of sociable impulses and exchanges which, for eighteenth-century Scotland, constitutes the definitive realm of humanity. Where, for other commentators tracing the formation in eighteenth-century Britain of 'normative' categories 'of the human', disability has been placed alongside racial and sexual difference as a further physical difference against which an emergent human 'norm' is defined, Blacklock's case suggests something quite different: the formation, in the Scottish Enlightenment, of notions of the human subject grounded not in the physical difference of the 'impaired' or 'normal' body, but on ideas of an essential human sociability common to all.[30] The force of Blacklock's equation, in his encyclopaedia article and his poems, of blindness with social isolation, however, asserts the difference between notions of common sociability in theory, and the possibility of socially significant performances of benevolence, compassion and sympathy by those excluded or marginalised by physical, class or other social difference in practice. Whilst, in theory, eighteenth-century Scotland finds no significance in such bodily differences as Blacklock's, beyond how they might provide an opportunity for compassion and sympathy, in practice its culture of sentimental sociability renders those such as Blacklock the passive objects of, rather than active participants in, its benevolent performances. A further irony, meanwhile, is presented by the ways in which the eighteenth-century Scottish sentimental culture which politely overlooks, 'cures' and compensates for Blacklock's physical disability, is itself deeply informed by and dependent on notions of the body, its impulses, physiological operation and the role of its senses, in ways in which the next chapter will explore. This emphasis on somatic sensibility was especially heightened in vitalist physiology, with its emphasis on automatic, unconscious response, and its language of bodily sympathy and sensitivity.

The four case studies considered in this chapter together demonstrate the variety of ways, across a range of eighteenth-century discourses, in which representation of the human subject drew on contemporary understandings of the body. All linked in some ways to the fertile enquiry into the nature of man which was a dominant concern of the Scottish Enlightenment, Arbuthnot, Armstrong, Blacklock and the moral philosophers demonstrate how thinking about the nature of the human self and identity was informed and extended by repeated references to the body in a number of overlapping and intersecting discourses – from theological, philosophical and literary, to poetic, critical and medical. Whilst the fascination with the nature of the body and its relationship to the mind is shared across these fields, such preoccupations produce different accounts and emphases in each writer's texts: from Arbuthnot's association of man's material embodiment with his fallen nature, to Armstrong's insistence on a model of governance and regulation, to Blacklock's assertion of a poetic identity independent of, yet undeniably moulded by, his inhabiting of an impaired body. The consistent recurrence of such themes and questions across Scottish philosophical and literary culture, widely construed, helps to explain how vitalism, with its new account of independent bodily operation, emerged in Scottish physiological theory at mid-century; it also suggests that such new developments would, as Chapters 2 and 3 demonstrate, be speedily assimilated into other, adjacent discourses.

2
Generating Sympathy: Sensibility, Animation and Vitality in Adam Smith and Mary Wollstonecraft

When Hume described the uncontrollable animation of Thomas Blacklock's body in a letter to Joseph Spence, as we saw in the previous chapter, this provided incontrovertible evidence of the blind poet's laudable 'sensibility': his taste, poetic discrimination and social sensitivity. A very different bodily activity (or lack of it) focuses Mary Wollstonecraft's depiction, in chapter 3 of her *Vindication of the Rights of Woman* (1792), of a woman whose purported sensibility connotes indolence, immorality and luxurious selfishness:

> I once knew a weak woman of fashion, who was more than commonly proud of her delicacy and sensibility. She thought a distinguishing taste and puny appetite the height of all human perfection, and acted accordingly. I have seen this weak sophisticated being neglect all the duties of life, yet recline with self-complacency on a sofa, and boast of her want of appetite as a proof of delicacy that extended to, or, perhaps, arose from, her exquisite sensibility; for it is difficult to render intelligible such ridiculous jargon. Yet, at the moment, I have seen her insult a worthy old gentlewoman, whom unexpected misfortunes had made dependent on her ostentatious bounty, and who, in better days, had claims on her gratitude. Is it possible that a human creature could have become such a weak and depraved being, if, like the Sybarites, dissolved in luxury, everything like virtue had not been worn away, or never impressed by precept, a poor substitute, it is true, for cultivation of mind, though it serves as a fence against vice?[1]

In this fierce attack on the fashionable but empty assumption of 'delicacy and sensibility', it is difficult to distinguish between Wollstonecraft's attitude to the various bodily habits and practices detailed here, and the moral corruptions which they are deemed to signal. The affected debilitations of the woman of sensibility's body are condemned as equally as the forms of moral impairment and decay which, for Wollstonecraft, they suggest, and the 'weakness' which is referred to three times in the short space of the quotation begins to speak not simply of the enervated body but, metonymically, of the woman's larger moral failings. The attack on the specific details of the 'weak woman's' physical lassitude – her reclining on the sofa, her lack of appetite, her self-proclaimed delicacy – thus readily extends to include her lack of moral fibre, a feebleness of virtue exemplified in the treatment of the 'worthy old woman', whose aged, and presumably weakened, body by contrast only underlines her own moral worthiness. The passage's mutual association of material and moral weakness culminates by explicitly expressing moral decay as physical deterioration – virtue has been 'dissolved' and 'worn away', human nature insufficiently 'impressed' with virtue or 'fenced' against vice. Wollstonecraft's metaphors here almost suggest that virtue has a material being, or physical embodiment, but her final evocation of the 'cultivation of the mind' gives her argument a different turn: the immaterial, intellectual and mental offers itself as remedy for a moral nature which has all too readily echoed the tendency to corruption of mankind's physical being.

Historians and critics of sensibility have long recognised the links between the eighteenth-century 'cult' of sensibility, as manifested in moral and aesthetic sensitivity, affected feeling and discriminating taste, and contemporary medical and scientific discourses of the body. In the wake of a pioneering article by George Rousseau, and in the context of an increasingly historically oriented criticism, critics have explored the connections between the culture of sensibility and a contemporary physiology of the nerves and senses, placing literary and other representations of the man or woman of sensibility in the context of developing knowledge of the nerves and senses traceable back to the research of Locke's teacher, Thomas Willis, in the second half of the seventeenth century.[2] Wollstonecraft's depiction of the 'weak woman of fashion', although a critique rather than a celebration of sensibility, and although problematically exhibiting the ease with which the bodily 'symptoms' of sensibility might be inauthentically reproduced, is clearly operating in just such a context as these commentators describe, in which sensibility, as a manifestation of social and moral impulses and instincts, is almost inseparable from a conception of a body whose ready, 'nervous' responses demonstrate and articulate its subject's sympathy and humanity.

This chapter, which explores the discourse of sensibility both in its origins, in Scottish Enlightenment moral philosophy, and in its reworking by Wollstonecraft at the end of the century, shares that interest in the relationship between contemporary accounts of the body and the theorisation of sensibility, demonstrating how Smith's description of the operation of sympathy, especially in the early stages of *The Theory of Moral Sentiments* (1759), draws on existing understandings of the body and the mind–body relation. This parallel between sensibility and contemporary physiology already in itself raises the question – hitherto unexplored by critics and historians – of the relation between sensibility and specifically vitalist physiology. These discourses have much in common, not least their mutual emergence in the mid-eighteenth century and their influential early theorisation by Scottish philosophers. Most particularly, however, sensibility shares with (or borrows from) vitalist physiology a body narrative which emphasises the same forms of automatic, unconscious physical response and independent somatic organisation as are foregrounded in, and essential to, sensibility. Sensibility shares with vitalist physiology an emphasis on nervous response, instinct and reflex, and addresses the same mysterious terrain of unconscious and immediate bodily activity which lies beyond the regulation and control of the conscious mind. All these parallels show how difficult, even impossible, it would be to theorise sensibility in relation to the mechanically based physiology which dominated in the early years of the eighteenth century.

As this chapter shows, however, the interconnection between sensibility (and the sympathy which is a crucial part of it) and dominant discourses of the body, such as vitalist physiology, is never straightforward. Sympathy, and sensibility, whilst dependent on empirical physiological and philosophical models of the body, which foreground the senses, impressions and nerves, must also necessarily disrupt and disturb those body narratives. Such disturbances enable the description of sympathetic feeling as a form of automatic response which (especially as first described) manifests itself through the body but is not simply of it. Equally – in another challenging conceptual problem – sympathy must be understood to be both an expected, usual and widespread response in our natures, and – as in the extraordinary moral animation demonstrated by Blacklock – something exceptional. It is both a function of our nature, and more than mere moral machinery: an operation, act or response whose cause both can be lengthily expounded in the extended account of who we are which constitutes Smith's *Theory*, and, unless we are simply sympathy machines, must always be something more than that.

The need to theorise sympathy in a way which not only shows how it originates in the human subject, as part of our natures, but also avoids a reductive moral mechanism points to perhaps the most fundamental shared concern between sensibility and vitalism. Both are ultimately concerned with the conceptual problem of generation – of how an essential and independent principle, whether of life or of moral feeling – originates, is sustained and operates. For many Scottish moral philosophers prior to Smith, the attempt to theorise a principle of social feeling relied uneasily on two alternative strategies, sometimes simultaneously deployed: either, like Francis Hutcheson, to suggest a 'moral sense' analogous to bodily senses, and thus risk moral determinism, or to fall back on traditional exhortation to virtue as an expected social and religious good, one for which humanity was designed by the divine creator. Smith's innovation in moral philosophy was to theorise a way in which a principle of moral feeling independently generates and sustains itself within human society, operating without the intervention of formal moral teaching or inculcation, and sufficiently systematically, without being a reductive mechanism. Smith's principle of sympathetic social feeling, nurtured 'within the breast' but often operating un- or semi-consciously, is shown to be a powerful force of moral animation: it informs human actions in every sphere of engagement, ensures the repetition and extension of actions which are socially recognised as morally good, and generally supports a self-sustaining system of morally animated subjects. As with vitalist physiology, Smith's theory of moral sentiments offers an independent, self-sustaining, self-regulating system which generates its own (moral) animation. As an independent, self-regulating, self-preserving system which produces the action necessary to its continual self-replication, this moral system is analogous too to Smith's later theorisation of the system of political economy, whose own parallels and shared origins with vitalist physiology are discussed in the next chapter.

All this means that Smith's extended theorisation of sympathy shares more than simply a body narrative with contemporary vitalist physiology. It also shares a theorisation of an inward power of active response whose activity defines and sustains the larger organisation to which it belongs. In vitalism, the active principle of life is ultimately mysterious, hidden in a mysterious terrain between mind and body, matter and spirit. In Smith, the active principle of sympathy is also perhaps best known through its observed effects, but the seemingly unanswerable, epistemological concern of vitalism – how the active spark which animates the whole can be generated – is in Smith answered in his account of moral feeling generated and replicated between moral spectators and subjects. In Wollstonecraft's writings at the end of the century, however, the generation of sympathy becomes altogether more uncertain.

As becomes evident too in Wollstonecraft's later critique of an ungenerative torpor of negative or absent feeling, sensibility thus operates in close relation with contingent discourses of bodily animation and vital generation, to involve sensibility, imagination and sympathy in the same complex questions of origin, generation and sustainability as are at the heart of vitalist concerns. Whilst Smith's account of sympathy begins with close attention to the body, he secures his account of animated moral feeling, as we shall see, by increasingly emphasising the operation of moral sensibility and the sympathetic body in culture. For him, moral response is closely associated with the sense of sight, and a distinction between sight and acculturated vision ensures that sympathy is part of our socialised natures without being a crude physical functioning. The variables of history and culture thus mediate and moderate what might otherwise be an overly determined science of moral man.

Meanwhile, Wollstonecraft, in her *Letters Written During a Short Residence in Sweden, Norway and Denmark* (1796), grapples with a paradoxical conception of sensibility as both self-generated and a product of imagination's active principles. Isolated in her lonely travels on the edge of Europe, meditating her abandonment by her lover and the collapse of French revolutionary ideals, sensibility for Wollstonecraft emerges as something never simply producible, yet whose appearance is always sought, an affectual response both dependent on the particular moment or circumstances of its generation, and, in its transcendence of that, capable of erasing mere genealogy, origin or history. Smith's concern, in expounding the operation of man's 'vital' social principle of moral animation, was to counter claims of our essential selfishness; his description of man's moral nature as independently and disinterestedly concerned for others produces us as fundamentally socially, not selfishly, oriented. Given the historical context of Wollstonecraft's text, meanwhile, the possibility of the re-emergence of sensibility's transformative modes of feeling brings with it the promise not just of personal recovery but the renewal too of ideals of political progress. Vitally present, powerful, animating and yet (especially for Wollstonecraft) mysterious, sensibility is thus as pivotal, within accounts of man's moral and social nature, as the understanding of the essential 'animation' of life was to contemporary philosophers of man's material being.

Corporeality and sensibility in early Wollstonecraft

The fundamental failing of the 'weak woman of fashion' in Wollstonecraft's early *Vindication of the Rights of Woman*, we have seen, is that her sensibility is confined to a bodily performance, expressed merely as a faked corporeal sensitivity or material delicacy which does not extend to her mind. If Blacklock's sensibility was expressed by his body, hers is limited to it; like Eliza, the mother of Mary in Wollstonecraft's first novel, another portrait informed by the same original source, she is a 'mere machine'.[3] Such an account of women as problematically confined, through a lack of education and hence intellectual development, to the corporeal, runs throughout the *Vindication*, whilst the opening chapters of *Mary* (1788), describe the kinds of necessary female mental and spiritual development which *Vindication* sees as generally so lacking. Indeed, although Wollstonecraft has been much discussed in the context of a 1790s 'radical sensibility', in which the subject's sublime self-elevation transcends or refuses a petty distinction between mind and body, such discussion should not cause us to overlook the striking presence, in these earlier texts, of an analysis in which mind and body, the intellectual and the physical, appear as frequently and straightforwardly opposed terms.[4] The *Vindication* diagnoses women's predicament largely as their limitation to the physical realm and the impeding of their development as creatures of reason. It repeatedly depicts them as slaves to their senses, unable rationally to entertain themselves or to fulfil their duties as mothers, and as using their bodies to attain a vicious power over their husbands and other men. For Wollstonecraft, only the acquisition of 'reason, virtue, and knowledge' will enable women to achieve 'the perfection of our nature', and the book's informing vision of such a possibility is at times expressed as an abandonment of the restraints of a dull material nature for a transcendent immateriality: 'an immortal soul, not restrained by mechanical laws and struggling to free itself from the shackles of matter, contributes to, instead of disturbing, the order of creation, when, co-operating with the Father of spirits, it tries to govern itself by the invariable rule that, in a degree, before which our imagination faints, regulates the universe'.[5] If a 'Romantic' urge for transcendence is readily identifiable in this heartfelt plea, it is notable too that such impulses are expressed in language which looks back to earlier expressions of corrupted corporeality and virtuous immateriality: John Arbuthnot's yearning desire for a mankind caught in the earthly realm to return to his heavenly 'native seat', for instance, might also be heard behind Wollstonecraft's words here. Wollstonecraft's articulation of the struggle exhorted by the *Vindication* as the rejection of a problematic corporeality by an immaterial soul – as the refusal by an active mind of an inanimate and bounded nature – reiterates an older language of mind and body even in her attempt to free herself from it. It also sketches the central problem for women as one concerned – like vitalism – with questions of self-generation and self-production.

For the Wollstonecraft of the *Vindication*, the body is central to such questions. It constitutes the 'shackles' of the grossly material and unanimated which constrain the female subject; but at the same time, a parallel insistence on physical activity and exercise also constructs the body, paradoxically, as a means of transformation, and identifies 'effort' and 'struggle' as productive of a looked-for redemption. The body itself, it seems, may have an internal force capable of redemptive self-generation. If Wollstonecraft's plea for a release from the 'shackles of matter' recalls an older tradition of religious yearning for immortality, the opposition between a dangerous indolence on the one hand and an energising and beneficial exercise on the other also replays familiar eighteenth-century attitudes to the body, such as those of John Armstrong considered in the previous chapter. Wollstonecraft's replaying of these themes produces the particular strenuousness of her writing, in which the georgic valuation of labour and exercise informs both a redemptive attitude to the body, and her own writing style. The moral value placed on 'exercise' – the insistence on deliberate and determined bodily activity – insists on finding a route to a higher mode of being within the 'shackles of matter' themselves. But whilst effortfulness is valued in Wollstonecraft, the problem often appears to be that of sustaining it. Impulses to such strenuousness inform her later fictional portrait of a 'woman of sensibility', Maria in *The Wrongs of Woman* (1798), for whom, unlike Eliza or the woman of fashion, sensibility entails a kind of constant work, a mental labour of preoccupation, meditation and imaginative anticipation – evident in, for instance, the maternal anxieties focused on her absent daughter. The Wollstonecraft of *A Short Residence in Sweden*, meanwhile, takes frequent refuge in bursts of activity as a release from the tensions and frustrations of her meditations – as though such physical outbursts might also provide the resolution sought by her labouring imagination. And Wollstonecraft's writing itself in *Vindication*, with its often ungainly but effortful repetitions, exclamations and expostulations, makes a kind of self-justified style from its undisguised striving. '[A]nimated' by her convictions, content with a 'rough sketch', Wollstonecraft disdains to 'cull my phrases or polish my style', and rejects the 'pretty superlatives' and 'sickly delicacy' of sentimental writing. She insists on her writing as labour, characterising it as both 'useful' and as an 'employment' for herself, and looks for it to prompt the same active force in her readers as it manifests in herself. Rather than offer writing as yet another luxurious product, as over-refined and ungenerative matter to be consumed by those of vitiated literary tastes, Wollstonecraft looks beyond mere words to the 'things' on which she hopes the work of her writing will produce some change, and in doing so effectively casts her writing as a productive tool, a necessary mechanism by which her labour might 'work' some reanimating transformation of the world.[6]

To an extent then, such activities of effort and exercise (in the body, in writing) promise a solution to the various restraints and repressions faced by women which the *Vindication* identifies. The physical exercise, and the striving towards reason, education and cognition with which it is associated and which are similarly exhorted, may ultimately prove transformative. But Wollstonecraft's commitment to the language and action of generation is fully revealed in her attack on physical lassitude and bodily torpor. The sterility and generative failure which they connote is strikingly expressed in the language of physical reproduction:

> All [women's] thoughts turn on things calculated to excite emotion; and feeling, when they should reason, their conduct is unstable, and their opinions are wavering – not the wavering produced by deliberation or progressive views, but by contradictory emotions. By fits and starts they are warm in many pursuits; yet this warmth, never concentrated into perseverance, soon exhausts itself; exhaled by its own heat, or meeting with some other fleeting passion, to which reason has never given any specific gravity, neutrality ensues.[7]

What John Whale has characterised as the 'strangely materialistic discourse of sensibility' is here peculiarly skewed towards the imagery of failed generation, the unsustained animal 'warmth' or infertility which contemporaries associated with female luxuriousness.[8] The failure to sufficiently form or nurture their sentiments into 'weighty matter' presents women's mental limitations here as repeatedly aborted gestations, or miscarried reasonings – a lack of vital spark or failed self-generation. This failure only emerges more strongly by comparison with a favourite figure of Wollstonecraft's, of reason 'warmed' by the heat of feeling to give birth to virtue:

> Sacred be the feelings of the heart! Concentrated in a glowing flame, they become the sun of life; and, without this invigorating impregnation,

reason would probably lie in helpless inactivity, and never bring forth her only legitimate offspring – virtue.[9]

If female sterility is the keynote of the previous passage, here Wollstonecraft offers an image of generation which resists conventional gender terms to emphasise the generative power of female feeling. Reason, conventionally masculine, is here classed as female; her 'bringing forth' is achieved through the concentrated power of heated feeling. Although called an 'impregnation', virtue's birth here is less the result of the passive female incubation of an active male principle, as in many contemporary accounts of generation, and more the product of precisely the female self-generative powers as are seen to be lacking in the earlier passage. Wollstonecraft's association of heat, itself strongly associated with the generation of life in vitalist accounts of conception, with female feeling, enables a new sketch of regenerative powers to be offered. Such repeated and exhortative phrases, like the language of corporeal and mental effort considered above, articulate Wollstonecraft's desire to point forward to what might be possible beyond current human states and reasonings, to new 'perfectible' possibilities, but they draw on a different version of corporeality – the body's capacity not to labour, but to reproduce – in doing so. Their language of generation is specifically bodily, and resonates with contemporary theories of the origins of life. This new rhetoric of animation, impregnation and vital creation explains what is lost in the failed generation of 'over exercised sensibility': possible 'new births' or newly animated being, such, perhaps, as are achieved in Wollstonecraft's novel *Mary*, announced in its 'Advertisement' as 'animated' by the 'soul' of the author, and sustained 'by a vivifying principle'.[10] Wollstonecraft's commitment to transcendence and making new, an urge or desire which elsewhere manifests itself by a turn away from the problematic materiality of the corporeal, is thus also articulated by recouping the vital possibilities of the body's own regenerative capacities. Such bodily possibilities also, as we shall see, provide a point of resolution for Wollstonecraft's later grapplings with the generation of sympathy in her *Short Residence in Sweden*. Meanwhile, in a very different text from Wollstonecraft's, Smith's *Theory of Moral Sentiments* secures its own account of sensibility's vital animation by locating the body in culture, so that human sociability becomes its own generative principle.

Sympathy and the senses in Smith's *Theory of Moral Sentiments*

Adam Smith is both an acknowledged and unacknowledged presence in Wollstonecraft's writings. He is deferred to as an authority worth quoting at length in the *Vindication*'s account of our natural sympathy with the rich, but anonymously present, and fiercely countered, in *A Short Residence*'s critical remarks on political economy. For Wollstonecraft, Smith is an important thinker whose writings – unlike many other sources which are mocked and derided – are to be taken seriously, whether or not they are in agreement with her own. Like any commentator on sensibility, on this topic too Wollstonecraft was working inevitably in Smith's wake, given that his *Theory of Moral Sentiments* (1759), an out-working of lectures given on moral philosophy at Glasgow University, was a foundational and exhaustive account of the operation of moral sensibility. Smith's moral philosophy – like Wollstonecraft's writing – offers an account of sensibility in which the body, and the relationship between the mind and the body, are unexpectedly important. Smith's detailed, even exhaustive, exposition of the operation of sympathy, the key manifestation of sensibility, depends on a closely mapped understanding of the mutual working of mind and body, which is both complex and productively ambiguous. Initially almost comprehensible as a near-physical function, sympathy at first seems firmly located within an account of human responsiveness in which mental perception and cognition are followed by bodily reaction. Smith's later elaboration of the role of imagination in sympathy, however, replays and reworks this story of mind and body to offer sympathy as a mode of moral animation which might both be independent of the body, and wreak at times uncanny effects on it. Elaborating a theory in which sympathy, the central social impulse, productively disturbs one's relation to oneself by making another present there – whether it be the object of one's sympathy, or the internalised impartial spectator, the 'man within the breast' – Smith describes our capacity to generate animated moral feeling only by disrupting an established account of the very working of our natures. In this, he reorients moral philosophy's close association with the empirical narratives of the natural philosophy, to emphasise the surprisingly generative capacities of mankind's moral being. Where earlier moral teachers relied on the 'external force' of exhortation to motivate virtue, Smith, like the vitalists, theorises a system in which the moral subject, in the principle of sympathy, generates its own animating principle. Like the 'vital principle' which ensures the life of the physical body, sympathy thus operates as the central animation of our moral and social natures, a manifestation of moral vitality powerful enough – as Hume witnessed with Blacklock – to disturb and refigure the otherwise straightforward functioning of our minds and bodies.

In making sympathy the central sign of man's sociability, Smith was in line with existing thinking in contemporary Scottish thought. The term 'sympathy', denoting compassionate fellow-feeling, already had a certain currency in early eighteenth-century moral philosophy, although it was not yet deployed or defined with any great precision. Smith's innovation in relation to 'a theory of moral sentiments' was to analyse the operation of man's compassionate feelings for others in great detail, through innumerable examples of their manifestation in a seemingly infinite range of social occasions or scenarios. But he also, significantly, theorised an account of sympathy as a morally generative social force in a way which supplemented existing accounts of man's physical, mental and perceptual faculties. A traditional moral philosophical approach to the emotions – which since the end of the seventeenth century had described the moral dangers or value of the various 'passions', and the need for their control – was thus supplemented with a more exact language of man's nature and constitution.[11] And the moral system Smith described detailed not a passive moral mechanism of force and counterforce, but one which had at its centre the generating and animating social force of sympathy.

Smith's description of mankind's moral nature, in keeping with the 'science of man' to which it contributes, is methodologically grounded in empirical description, yet whilst the vital sentiment of sympathy is observationally present, certain residual theoretical lacunae – which parallel those in theories of vitalism – remain. On the one hand, Smith offers a newly precise description of man's moral being, in which a Lockean 'natural history' of cognition – of sense impressions producing mental perceptions and ideas – renders in newly 'scientific' light the matter of mankind's moral nature and social relations. On the other hand, however, an ambiguity which remains in Smith's account of the moral sentiments, and especially about their status as mental 'feeling' or physical 'sensation', works to blur the significance of that distinction between the mind and the body, and begins to articulate the possibility of a 'sentiment' which is an experience of, and experienced by, both. Whilst an essentially empirical model of human nature founds Smith's account of the moral sentiments, sympathy and the imagination, the attention to a mode of experience which is neither limited to nor contained by either the mind or the body undermines the very opposition between mind and body which such an empirical model deploys, and enables the emergence of an alternative vision of an integrated, 'embodied' moral being whose sympathies, sentiments and impulses mutually animate both the mind and the body.

Since Hobbes' assertion of mankind's natural selfishness, and Bernard Mandeville's provocative demonstration, in *The Fable of the Bees* (1714), of mankind's virtue as fundamentally self-interested, a number of philosophers had attempted to demonstrate the existence of a disinterested sociability in human nature. In such attempts, allying a sympathetic moral response with man's constitution – rendering it as 'natural', or analogous to the body – was both tempting and problematic. Whilst it might strengthen the case for what many saw as a fundamental humanity in man, it also risked presenting such moral instincts as mere machinery. The suggestion by Hutcheson, an important influence on Smith, of the existence in man of a 'moral sense', a more systematic presentation of Shaftesbury's insistence on man's innate response to beauty and virtue, illustrated these very problems. Hutcheson's description of the moral sense as analogous to the five external senses of the body was criticised by Smith in *The Theory of Moral Sentiments*, where he argued that if such a moral sense existed, it would 'judge with more accuracy' a man's own conduct than that of others, 'of which it had only a more distant prospect', a possibility which is rarely experienced.[12] Other philosophers offered their own accounts of mankind's sociable instincts. For Archibald Campbell, Professor of Divinity and Church History at St Andrews University from 1731, self-love, combined with an identification with others based on our resemblance to them, was the foundation of social sympathy. 'Man naturally likes his Kind, as they bear his Likeness and Image,' he asserted, and quoted from Dryden's translation of a Juvenal satire to found the difference between the human and the bestial on this natural fellow-feeling between mankind:

> Compassion proper to Mankind appears,
> Which Nature witness'd when she lent us Tears.
> Of tender Sentiments we only give
> Those Proofs: To weep is our Prerogative;
> To shew by pitying Looks, and melting Eyes,
> How with a suff'ring Friend we sympathize
> ...
> *Who can all sense of others Ills escape,*
> *Is but a Brute at best in human Shape.*

But beyond asserting that 'the human Constitution lays every Man inevitably open to *Sympathy* and Fellow-feeling', Campbell goes no further in elaborating the detailed operation of sympathy.[13] David Fordyce's *Elements of Moral Philosophy* (1754) describes the passions as 'contagions', which, perhaps aided by an eloquent voice or countenance, are transmitted 'like a subtle Flame, into the Hearts of others', to raise 'correspondent Feelings there'. Whilst strongly asserting the existence of, in Fordyce's words, 'a peculiar and strong Propensity in [man's] Nature to be affected with the Sentiments and Dispositions of others', the formulations of each of these philosophers run the risk of either, as in Campbell's emphasis on self-love, founding sympathy on selfishness, or, as in Hutcheson's assertion of a 'sense' or instinct, reducing sympathy to become merely a determined function of man's nature, and thus stripping it of any moral significance or active choice.[14] The secure description of sympathy as a disinterested but integral part of human nature had thus to avoid identifying it with self-interest on the one hand, or seeing it as produced by a crude 'mechanism' of human nature on the other.

In Smith's account of the moral sentiments, this dilemma leads to a careful account of man's 'nature' or 'constitution', especially evident in the negotiation of the relationship between sympathy and the body. Too closely associated, and, like Hutcheson, he would run the risk of presenting man's moral judgement as mere physical determination; at the same time, the 'science of man' demanded an account of man's moral 'nature' which went beyond the unelaborated assumption, as in Fordyce's *Elements*, of some 'Pieces of Machinery' which governed man's sociable impulses.[15] In his own account of sympathy and 'fellow-feeling', Smith deployed a technical vocabulary of the senses, impressions and sensations, to initially align his exposition of the operation of sympathy with accounts of sensation and perception in empirical philosophy, but he also disturbed the relationship between the subject and his body through the intervention of the imagination, which becomes the key faculty in the subject's sympathetic feeling for another. The effect, as in Smith's first extended example of the operation of sympathy, is the offering of a natural history of 'fellow-feeling', a genealogy of sympathetic experience and a construction of the sentimental subject which also disturbs empiricism's received model of the 'self', and the subject's relations with others:

> [W]e can form no idea of the manner in which [other men] are affected, but by conceiving what we ourselves should feel in the like situation. Though our brother is upon the rack, as long as we ourselves are at our ease, our senses will never inform us of what he suffers. They never did, and never can, carry us beyond our own person, and it is by the imagination only that we can form any conception of what are his sensations. Neither can that faculty help us to this any other way, than by representing to us what would be our own, if we were in his case. It is the impressions of our own senses only, not those of his, which our imaginations copy. By the imagination we place ourselves in his situation, we conceive ourselves enduring all the same torments, we enter as it were into his body, and become in some measure the same person with him, and thence form some idea of his sensations, and even feel something which, though weaker in degree, is not altogether unlike them. His agonies, when they are thus brought home to ourselves, when we have thus adopted and made them our own, begin at last to affect us, and we then tremble and shudder at the thought of what he feels. For as to be in pain or distress of any kind excites the most excessive sorrow, so to conceive or to imagine that we are in it, excites some degree of the same emotion, in proportion to the vivacity or dulness of the conception.[16]

In this complex and detailed picture, the experience of sympathy is one in which the body is both central and strangely displaced. It is the experiences of another's body – our brother on the rack – which prompts sympathy, conceived as an imaginative 'copying' of physical experience. The body isolates us from others – our senses alone 'will never inform us' what they feel, they 'never did and never can' take us beyond ourselves – but it also, through the intervention of the imagination, connects us to them, through a physical replaying of their sensations. Sympathy, an act of imagination, produces a replication of the senses of another's body, or, more precisely, a hypothetical representation of what our own senses would be, were we to experience what 'our brother' does. This act of imagination thus reorders not only our usual relationship to our senses, but what our senses signify: rather than offering direct experience, our senses offer a hypothesised sensation, a possible, fancied suffering, which is at once ours and the experience of another, because through it we 'enter into his body', and 'even feel' something like what he feels. Sympathy is thus, in Smith's first examination of it, the strange replication of the senses of others in our own bodies, an imaginative repetition or mimicking of others' bodies by our own: a very material, if ultimately mysterious, means by which our sociable relation to them is secured.

Here Smith's vision of the 'machinery' of sympathy operates in close relation to established accounts of man's perceptual and physiological make-up. Yet it also leaves some central questions unanswered. The account describes a close relationship between the mind and the body – the imagination is able to 'represent' physical sensation, and elicits a physical reaction in our 'trembling' and 'shuddering' – but it also poses more questions than it answers: how *can* the imagination intervene in the body like this? How can the body be affected by a representation of suffering? The relationship between mind and body is central but ambiguous: a confusion evident throughout the literature of sympathy, from pre-Smithian moral philosophy's inability to define the mechanism of sympathy as a mental or physical operation, to medical accounts of pathologised sensibility – hysteria, the vapours, spleen and so on – as having both physical and mental causes and symptoms.[17] It could even be seen in the only account of sympathy to rival Smith's in detail and thoroughness: David Hume's account of sympathy, in his *Treatise*, describes it as a 'contagion' of passions, in which the ideas of the mind become, in some unexplained way, impressions of the body, a reversal of the usual transformation of impressions to ideas.[18] Interestingly, Hume backed away from precise explanations of the mental and physical mechanisms of sympathy in his later recastings of the *Treatise*: his *Enquiry Concerning Human Understanding* characterised the relationship between mind and body as 'something still more minute and more unknown', and as 'mysterious and unintelligible', and his *Enquiry Concerning the Principles of Morals* was content simply to note that a 'principle' of 'humanity or fellow-feeling' existed.[19] Thus whilst Smith's exposition of sympathy succeeds in linking it with the particular somatic signals – sighing, trembling, fainting – which were to mark the social enactment of sensibility, its assertion of the physical manifestations of sympathy masked a deeper ambiguity about the physical and psychological functions by which the experience was produced – even whilst insisting on its capacity to animate the body in new and dramatic ways. Sympathy, Smith's 'vital principle' of social feeling, is, like the vitalist's animating force, located in the same unknown point of interplay between mind and body. At the heart of Smith's exposition of sympathy lies the same mystery of bodily action which vitalist physiology also acknowledged.

As though in recognition of this foundational, but unacknowledged, mystery, sympathy can often be seen to operate in strange or unaccountable ways, so that the sociability it announces arrives via a disturbing disjunction of the subject's pristine self-sufficiency. As we have seen, sociability is marked by a displacement of one's relation to oneself, as the 'other', the object of sympathy, enters into our own body through the senses' iteration of what he or she may be feeling. The physical responses regarded as markers of sympathetic sociability – tears, shuddering – are thus, as Smith's exposition makes clear, more precisely signs of the internal generation of an imagined, 'spectral' or phantasmatic representation of the object of sympathy, whose experience directly informs the body. This explains the often uncanny-seeming manifestations of sympathy, identifiable already in Hume's disturbing description of mental ideas becoming physical impressions, a reversal of the usual empirical account of the perceptual process as sense impressions producing ideas, and which, as in the Freudian uncanny, blurs the distinction between the real and the imagined. Such a disturbing blurring, between the real and its ghostly effects, between the subject's 'own' experience and the sympathetic traces of that of others, is most marked in Smith's discussion of our sympathy with the dead, in which we lodge 'our own living souls in their inanimated [*sic*] bodies', or in our sympathy with a just revenger, when we 'enter, as it were, into his body, and in our imaginations, in some measure, animate anew the deformed and mangled carcass of the slain'.[20] The weirdness of these examples, with their reanimation of the dead by the living, in turn throws into relief the ways in which, in some of Smith's other examples, sympathy appears as the 'haunting' of bodies by the physical sensations of others, a strange transfer of a moment of another's material experience:

> When we see a stroke aimed and just ready to fall upon the leg or arm of another person, we naturally shrink and draw back our own leg or our own arm; and when it does fall, we feel it in some measure, and are hurt by it as well as the sufferer. The mob, when they are gazing at a dancer on the slack rope, naturally writhe and twist and balance their own bodies as they see him do ... Persons of delicate fibres and a weak constitution of body complain, that in looking on the sores and ulcers which are exposed by beggars in the streets, they are apt to feel an itching or uneasy sensation in the correspondent part of their own bodies.[21]

Sympathy here appears as much being infected, as affected, by the sufferings of others, not through a physical 'contagion' but through an imaginative 'copying' of their experience, a displacement of one's own for another's experience which is the nature, even the price, of sociability. Whilst Smith's reference to 'persons of delicate fibres and a weak constitution' might suggest that sympathy is simply a kind of physical weakness – perhaps, in Wollstonecraftian terms, to be combated with increased vigour and exercise – what such examples illustrate is not so much the experience of a particular kind of body, perhaps characterised by the peculiarly sensitive or delicate nerves so vaunted in the 'cult' of sensibility, but rather how, in sentimental culture, one's usual embodiment of one's physical self is vulnerable to being displaced by the actions of the sympathetic imagination, and their physical consequences.[22]

The disruption of the subject's usual relation to his or her body in acts of sympathy, evident in these examples, is extended further as Smith's discussion increasingly emphasises the role of the imagination. Whilst his early examples usually demonstrate a sympathy with physical suffering – our brother on the rack, beggars' ulcers, sore eyes, an ill baby – and hence emphasise the subject's imagined participation in that physical pain, the discussion of sympathy increasingly emphasises imagination rather than any specifics of bodily sensation. This emphasis on imagination is extended to generate an entire theory of sentimental sociability, so that sympathy and social identity are mutually subject to and produced by the imagination. The subject's imaginative relation to others, and theirs to him or her, as well as, crucially, the imaginative construct of the 'impartial spectator', defines and governs the proper operation of sympathy, so the enactment of sympathetic sociability, although perhaps marked by physical signs and symptoms, is always an act of imagination. In fact, it is only by taking up a position in relation to one's imagined projection of what others' imaginations of you might be, or, more complexly, in relation to your projection of the opinions of the 'impartial spectator', the fantasised ideal being who defines a standard of proper action and behaviour, that the subject assumes his proper place in sentimental society. For Peter de Bolla, this means that the *Moral Sentiments* offers a theory of spectatorial subjectivity, in which the subject is produced by being inserted into a field of vision, or certain conditions of visibility and visuality, but given that the sentimental subject's most significant relations are not those of sight, but those of imagination – the 'impartial spectator' is clearly an imagined being, and imagination clearly tempers and governs all acts of vision described by Smith – the book might equally be described as the production of the subject within a collectively produced imaginative field.[23] Perhaps another way of saying this, as even Smith's first example of the brother on the rack makes clear, is that seeing is also imagining, an observation which hints at the unlimited extent of the imagined, and its disturbing dislocation of what might be 'real'. One way of reading Smith's more focused attention to the function of the imagination, discussed in the next section of this chapter, is as an attempt to head off such possibilities by presenting the imagination as simply another form of sight – albeit one whose identity as an acculturated power is readily evident.

Imagination, vision and moral animation

The relationship between sight and imagination is central to Smith's account of the generation of sympathetic moral response: our moral vitality is visually founded. Imagination's capacity to generate moral response is rooted in the visual sense of the body, whilst its possibilities extend towards the ultimately unattainable moral perfection of the impartial spectator. *The Theory of Moral Sentiments* is thus deeply interested in mankind's visual capacity and the responses our sight trigger in us, but develops its account of the moral sentiments not as an automatic function of vision, but as dependent on an acquired ability to see in a particular way. Smith develops his concern with sight as the text progresses to present a mode of vision which is in part imaginative, and which, unlike ordinary sight, is central to the process of animating the subject to new processes of moral cognition. Imaginatively informed, culturally sensitised, sight thus generates the vital force of sympathy.

Given that almost every example of sympathy offered in *The Theory*, in endlessly elaborated instances and scenarios, involves the act of looking, it is not for nothing that the sympathiser is often called the 'onlooker', or that the moral sentiments, as Smith told his students, could be termed 'moral observations'.[24] The importance placed on sight throughout the work in part reflects that sense's significance in a mid-century society where the mutual display of self and observation of others constituted foundational acts of sociability – where, as others have noted, the key periodical of fashionable culture was called *The Spectator*. Smith's *Theory* adds to this sense of society as visual theatre by giving a moral weight to both acts of viewing and acts of performance, and by integrating the act of vision with that of moral judgement.[25] Visual apprehension and sympathetic response often appear nearly or actually simultaneous in Smith's text. This is very clearly so in those examples where the cause of suffering endured by the object of sympathy may be visually perceived: the sores and ulcers of beggars in the streets, the madman who is unaware of his tragic loss of reason, the 'tearing' of flesh in a body undergoing a surgical operation.[26] Even when his examples of the often physical sufferings of the body give way to a more 'polite' concern with the mental and emotional trials of others, rather than their physical distress, the imagination needed to perceive, understand and sympathise with such states and experiences is nevertheless presented as in many ways another form of vision, enabling sympathetic access to the emotions and internal moral dilemmas where precise acts of moral adjudication from the onlooker are Smith's real concern in this work. The importance of sight is evident even at the level of textual genealogy: Smith draws on an earlier, youthful essay on the operation of the 'external senses' to figure the operation of the imagination at a crucial moment, so that even when presenting a fully elaborated theory of the sympathetic imagination, a concern with the processes and function of sight remains present.

These links between sight, imagination, cognition and animation, which are so fundamental to Smith's account of the formation and operation of our moral sensibilities, are present too in other earlier works. His essays on 'The Principles which lead and direct philosophical enquiries; illustrated by the History of Astronomy' and 'The Principles which lead and direct philosophical enquiries; illustrated by the History of Ancient Physics', both carefully preserved from destruction at the end of his life when many of his other papers were deliberately destroyed, between them present a fully formed exposition of Smith's theory of philosophical progress, in which acts of looking, and acts of imagination, respectively present and solve problems of understanding and cognition. 'Philosophical enquiries', according to Smith, are stimulated by problems initially presented by vision: specifically, the disparity between seeing and understanding evident when we look at the 'theatre of nature' but cannot comprehend it. Nature's 'seeming chaos of dissimilar and disjointed appearances' presents a picture of a confusing and disturbing world, and makes evident the need to render it into a 'coherent spectacle to the imagination' by formulating the various 'qualities, operations, and laws of succession ... along which the imagination could glide smoothly, and without interruption'.[27] Stimulated by a visual problem – our inability to understand what we see – philosophy fittingly addresses its solution to that problem in visual terms, producing an explanation of the 'spectacle of nature' which enables the imagination to contemplate and smoothly connect what it sees with what it understands. The shift from sight to imagination, from actual sight to imagined vision, frames cognition in visual terms, as well as linking, or even blurring, actual and imagined forms of sight: from the moment in which 'philosophical enquiries' have begun to produce or concoct explanations for what is seen, looking is never divorced from a more interpretative, imaginative act. Entwined though those two acts of looking may be, however, their integration is often never complete or absolute, and indeed, as Smith's histories of astronomy and ancient physics are traced, the continued progress of philosophy is itself attributed to repeated disjunctions between what is seen and what is imagined to cause or produce such appearances: observations which do not fit established explanations prompt renewed enquiries.

In part, Smith's account of knowledge acquisition presents it as an aesthetic and psychologically informed process: philosophy satisfies our aesthetic and imaginative capacities, as much as our reasoning ones. To the satisfied imagination, a successfully linked chain of ideas almost appears to have a life of its own, seeming to 'float through the mind of [its] own accord', falling in 'with the natural career of the imagination' so that 'thought glides easily along them'.[28] Such a pleasurable animation of the mind accounts for what Smith describes as the pleasures of philosophical pursuits, but animated imaginations can present their own problems. The tendency of ancient peoples to understand 'whatever particular part of Nature excited the admiration of mankind' as being 'animated by some particular divinity; so the whole of Nature having, by their reasonings, become equally the object of admiration, was equally apprehended to be animated by a Universal Deity, to be itself a Divinity, an Animal', illustrates precisely such a problem: the beliefs of the 'rude ages of the world' fulfil imaginative and psychological needs as much as the more accurate physical theories of, say, Newton.[29] Such 'occult' beliefs, to use Newton's own term, pose the question of the relationship between sight and imagination in Smith's account of philosophy, and suggest the dangers of too much imagination and too little sight, a danger which the procedure of Smith's essays, which trace the increasing improvement in scientific theories from such ancient myths to contemporary discoveries, presents as increasingly diminished with historical 'progress'. Proper philosophy, Smith would assert, is about sight and imagination working harmoniously together, a conjunction of empirical observation and imaginative insight represented for him by Newton's achievements in his own age. Even Smith himself, who has been 'endeavouring to represent all philosophical systems as mere inventions of the imagination' has 'insensibly been drawn in' to describe the principles of Newton's system 'as if they were the real chains which Nature makes use of to bind together her several operations'.[30] Although, concluding his account of the history of astronomy with Newton, maker of 'the greatest discovery that ever was made by man', Smith shows how far that science has come since the crude beliefs of 'rude' nations, and thus fulfils his own argument about progress, his own language, which slips from describing Newton's discovery as a 'system' to nature's actuality, repeats the same tendency to 'believe' which he would otherwise limit to an earlier historical age. However much his essays might want to confine a ready admiration and belief to an earlier age, Smith's own admiration for Newton suggests such a historical difference might be less marked than he might wish: Smith commits the same act – of endowing his beliefs with 'animation' – as he attributes to the 'ruder' peoples.

The dangers of such 'occult' beliefs as those discussed in the 'History of Ancient Physics' nevertheless signal the importance of proper vision informing the imagination to generate and animate our perception of the world. *The Theory of Moral Sentiments*, as we shall see, works carefully to regulate the relationship between sight and vision, maximising the potential of the imagination to extend vision to what cannot be seen, but rooting it in the realities of sight, to give rise to a fully animated moral being. Smith's disparagement of 'rude' theories of animation should not mask the fact that it is precisely by such a language of 'animation' that he often articulates his own interest in the functioning of mankind's moral nature. The metaphor of the 'theatre of nature' used in the 'History of Ancient Physics' itself implies, with its distinction between appearances and whatever produces them, some hidden animating causes carefully located behind the scenes, and *The Theory of Moral Sentiments* similarly, in a discussion of the differences between means and ends in the moral operation of the mind, illustrates that distinction by drawing attention to the animating mechanisms of plant or animal life, as distinct from their ultimate causes.[31] The 'digestion of food, the circulation of the blood, and the secretion of the several juices which are drawn from it' – the 'efficient' causes of animal life – are easily distinguished from its final cause, Smith says, yet in the consideration of our own minds, efficient and final causes are readily confused: we fail to see an ultimate, divine cause in our own moral reasonings. The parallel with the investigation of animal or plant life demonstrates Smith's own determination to understand the 'efficient' functioning of mankind's own moral nature, a concern which runs together in the text with a more direct one: to 'cause' such moral animation, by inspiring his readers to actions which are 'generous and noble', an important purpose to be fulfilled by moral philosophy and one in which the casuists, for instance, fail.[32] The *Theory* wants both to trace the 'efficient' causes of our moral animation, and to produce it – to animate moral function in its readers just as the circulation of the blood or the digestion of food produces the vitality of organic life forms.

Smith's account of the kind of moral writing which will 'animate us to the practice of virtue' uses vision for its governing metaphor. Moral animation is the acquisition of a certain mode of seeing, and this means that even reading moral philosophy itself gets figured as a kind of visual perception. The best kind of moral philosophy, Smith asserts, will

> present us with agreeable and lively pictures of manners. By the vivacity of their descriptions they inflame our natural love of virtue, and increase our abhorrence of vice: by the justness as well as delicacy of their observations they may often help both to correct and to ascertain our natural sentiments with regard to the propriety of conduct, and suggesting many nice and delicate attentions, form us to a more exact justness of behaviour, than what, without such instruction, we should have been apt to think of.[33]

Moral writing here is a kind of painting, offering less a cognitive than an affecting and suggestive experience to the reader. The 'inflammation' of the reader's 'natural love of virtue' suggests an almost passive construction of the text's 'onlooker': all the 'vivacity' here lies with the text. But against that model of inevitable contagion, of passive response, lies the importance too of being able to 'read' or 'see' the text's picture properly, and respond with discrimination. A language of aesthetic evaluation and taste is drawn on here – the reader must discern the 'justness' and 'delicacy' of the text's 'observations' – as much as that of an inevitable 'animation'; in fact, although 'inflammation' might seem immediate, such language suggests that it follows a more careful evaluation. Whilst moral writing offers an education in the acquisition of judgements and ability to discriminate, like those exercised in matters of aesthetic taste, its being figured as mediating a kind of sight means that, whilst a form of 'inflammation', infection and animation, moral acquisition is also constructed as the attainment of an informed, educated mode of vision.

The same notion of the 'animation' of the moral sentiments through the education of vision is replayed repeatedly through the text. Vision's centrality to Smith's account of the 'efficient' animation of our moral sentiments is evident for instance in the description of recurring actual or imagined reciprocal gazes between an observer and an observed, visual exchanges which produce society as a visual field in which awareness of others' perspective modifies one's natural dependence on one's own view. At this most basic level, sight, and a rudimentary awareness of perspective, offers an education in moral perspective. Thus, when Smith suggests that if a man 'views himself in the light in which he is conscious that others will view him, he sees that to them he is but one of the multitude in no respect better than any other in it', the imagined exchange between viewer and viewed sketches a form of moral equivalence between social peers, and stimulates moral sensibilities.[34] This exposition of moral sentiments as animated by sight is developed in more extended discussions. Reiterating his theme of the need to view ourselves as others might, Smith offers the following:

> As to the eye of the body, objects appear great or small, not so much according to their real dimensions, as according to the nearness or distance of their situation; so do they likewise to what may be called the natural eye of the mind: and we remedy the defects of both these organs pretty much in the same manner. In my present situation an immense landscape of lawns, and woods, and distant mountains, seems to do no more than cover the little window which I write by, and to be out of all proportion less than the chamber in which I am sitting. I can form a just comparison between those great objects and the little objects around me, in no other way, than by transporting myself, at least in fancy, to a different station, from whence I can survey both at nearly equal distances, and thereby form some judgement of their real proportions. Habit and experience have taught me to do this so easily and so readily, that I am scarce sensible that I do it; and a man must be, in some measure, acquainted with the philosophy of vision, before he can be thoroughly convinced, how little those distant objects would appear to the eye, if the imagination, from a knowledge of their real magnitudes, did not swell and dilate them.

'In the same manner', Smith continues, our own interests always appear greater and more important than those of other people. The interests of another man,

> as long as they are surveyed from this station, can never be put into the balance with our own, can never restrain us from doing whatever may tend to promote our own, how ruinous soever to him. Before we make any proper comparison of those opposite interests, we must change our position. We must view them, neither from our own place nor yet from his, neither with our own eyes, not yet with his, but from the place and with the eyes of a third person, who has no particular connexion with either, and who judges with impartiality between us. Here, too, habit and experience have taught us to do this so easily and readily, that we are scarce sensible that we do it; and it requires, in this case too, some degree of reflection, and even of philosophy, to convince us, how little interest we should take in the greatest concerns of our neighbour, how little we should be affected by whatever relates to him, if the sense of propriety and justice did not correct the otherwise natural inequality of our sentiments.[35]

The 'philosophy of vision', the lessons of sight, teach us our true moral proportions, uncovering our 'real littleness' from the 'natural misrepresentations of self-love'.[36] A tightly plotted comparison with sight, and a learned acquisition of visual perspective, demonstrates how the imagination can be educated into acquiring a similar sense of moral perspective. The lessons learned from considering sight directly inform our capacity for imaginative or moral cognition, by modifying our untutored perspective with that of an actual or imagined other. The notion of an 'impartial spectator', however, whilst offered as an abstract, impersonal, transcendent and thus somehow more perfect embodiment of moral judgement, also reveals the limits of a parallel between visual and moral perception, as the ideal moral judgement represented by the impartial spectator, figured as a perfect vision, is also revealed as an ideal which is impossible to attain. Our moral evaluations are always to be made in the shadow of an ideal, perfect moral judgement which, whilst we might hope to attain, we cannot repeat. At the same time, the impartial spectator extends the operation of vision into an imaginative realm unrestrained by the viewer's physical location, offering the possibility of vision unshackled by the restraints of perspective. Moral feeling is animated by reference to the ideal, abstract and imaginative function of the impartial spectator: if his perfect moral vision cannot be attained, it can nevertheless be emulated.

The specificity of Smith's emphasis on the imaginative generation of moral sight is evident when contrasted with another prominent eighteenth-century account on the imagination: that given by Addison in his *Spectator* essays. Again taking his cue from philosophies of vision – this time Locke's assertion that light and colour are not properties of an object but 'ideas in the mind' – Addison offers a fantasia in which imagination adds new and wonderful visions to transform an otherwise bleak world:

> Things would make but a poor appearance to the eye, if we saw them only in their proper figures and motions ... We are everywhere entertained with pleasing shows and apparitions, we discover imaginary glories in the heavens and in the earth ... [O]ur souls are at present delightfully lost and bewildered in a pleasing delusion and we walk about like the enchanted hero of a romance who sees beautiful castles, woods, and meadows, and at the same time hears the warbling of birds and the purling of streams; but upon the finishing of some secret spell the fantastic scene breaks up, and the disconsolate knight finds himself on a barren heath or in a solitary desert.[37]

Here sight and imagination are vehicles of pleasing delusion, adding 'supernumerary ornaments to the universe' to produce a Quixotic landscape of romance. It is a very different scene from the more prosaic prospect of lawns, woods and mountains, a picture actually visible from his window, which founds Smith's account of sight's influence on the imagination, and its lesson about the relationship of sight to the imagination is proportionally dissimilar too. Whilst for Addison, our perception of beauty stimulates the imagination to produce new ideas in response to what is seen, a process Addison is happy to present as pleasurable fantasy, Smith modifies the Addisonian tendency to assimilate imagination to fantasia, bringing it back to the discipline of sight. For both, the possibilities of imagination depend, paradoxically, on what can be seen, but if imagination for Addison equates to the unfettered fancy of romance, for Smith it is a hypothesised vision, an extension of sight into what cannot be seen, which can in turn offer a greater vision, a more authoritative insight, into what appears to be present. For Addison, what gets animated, in a false, misleading way, is the imagination, which gives rise to an enjoyable but deluded romance of the self, who is imagined to be 'like the enchanted hero of a romance', at the centre of a world designed for his pleasure. What is animated in Smith through the imagination is precisely the opposite: not romantic and flattering fictions of ourselves, but a means of critical judgement capable of overturning exactly such misperceptions, puncturing our usual tendency to delusion with an authoritative counter-perspective. Imaginative animation for Smith here, as in his account of sympathy and sensibility more generally, gives life to moral awareness, not romantic excesses, as the imagination offers a new form of vision not to obscure or ornament what can be seen, but to deepen and extend it.

Sympathy and generation

In *The Theory of Moral Sentiments*, Smith finds, in imagination informed by sight, the active spark which generates moral sensibility. One narrative of how the acquisition of moral sense in this way produces a transformation of being, a new animation of the self, is given in his tale of a man brought into society after living in 'some solitary place, without any communication with his own species'. Such a man, he says, whilst living alone

> could no more think of his own character, of the propriety or demerit of his own sentiments and conduct ... than of the beauty or deformity of his own face. All these are objects which he cannot easily see, which naturally he does not look at, and with regard to which he is provided with no mirror which can present them to his view. Bring him into society, and he is immediately provided with the mirror which he wanted before. It is placed in the countenance and behaviour of those he lives with, which always mark when they enter into, and when they disapprove of his sentiments; and it is here that he first views the propriety and impropriety of his own passions, the beauty and deformity of his own mind. To a man who from his birth was a stranger to society, the objects of his passions, the external bodies which either pleased or hurt him, would occupy his whole attention. The passions themselves, the desires or aversions, the joys or sorrows, which those objects excited, though of all things the most immediately present to him, could scarce ever be the objects of his thoughts. The idea of them could never interest him so much as to call upon his attentive consideration. The consideration of his joy could in him excite no new joy, nor that of his sorrow any new sorrow, though the consideration of the causes of those passions might often excite both. Bring him into society, and all his own passions will immediately become the causes of new passions. He will observe that mankind approve of some of them, and are disgusted by others. He will be elevated in the one case, and cast down in the other; his desires and aversions, his joys and sorrows, will now often become the causes of new desires and new aversions, new joys and new sorrows: they will now, therefore, interest him deeply, and often call upon his most attentive consideration.[38]

An education through sight is the means of generating moral sensibility in the once-solitary man, but sight can only generate moral animation in a social context. Sensibility's generative principle is thus, in a complex formulation which defies the ostensible simplicity of this tale, the social operation of an imaginatively informed physical sense. When Wollstonecraft considers sympathy's generation in a very different text some half-century later, her solution will be very different to Smith's rooting of sympathy's 'active principle' in the Enlightenment values of education and sociability.

Smith's sense here of vision as more than the mere operation of a sense-organ is perhaps informed by an early essay on the 'external senses', which devotes most of its space to a discussion of vision. The essay quotes extensively from an account by the surgeon William Cheselden of a cataract removal operation he performed on a 12-year-old boy, and the patient's subsequent gradual 'mastery' of sight. Cheselden's operation had initially come to prominence because it had seemed to provide experimental evidence to answer a query considered by Locke, in his *Essay Concerning Human Understanding*, about the relationship between sight and touch. Cheselden's account of his patient's learning the connection between these two senses directly addresses this question, but Smith's interest, indicated by his pattern of quotation from Cheselden's case-history, appears to lie in the detailed description of vision as an acquired perceptual capacity:

> When he [the patient] first saw ... he was so far from making any judgment about distances, that he thought all objects whatever touched his eyes (as he expressed it) as what he felt did his skin; and thought no objects so agreeable as those which were smooth and regular, though he could form no judgment of their shape, or guess what it was in any object that was pleasing to him. He knew not the shape of any thing, nor any one thing from another, however different in shape and magnitude; but upon being told what things were, whose form he before knew from feeling, he would carefully observe, that he might know them again; but having too many objects to learn at once, he forgot many of them; and (as he said) at first learned to know, and again forgot a thousand things in a day. One particular only (though it may appear trifling) I will relate: Having often forgot which was the cat, and which the dog, he was ashamed to ask; but catching the cat (which he knew by feeling) he was observed to look at her stedfastly, and then setting her down, said, So, puss! I shall know you another time.[39]

Smith's account culminates with the description of the patient's viewing of 'a large prospect' from the Epsom Downs, where he exclaims over a 'new kind of seeing', and it is only now that he is declared by Smith to be 'completely master of the language of Vision'.[40] Such pronouncements suggest that the 'young gentleman' has progressed beyond a sense of sight confined to a mechanical identification of particulars, to something quite different: he is now not merely viewing the objects available to his sight, but perceiving the beauty of a 'prospect' valued in aesthetic terms by his social peers. His acquisition of sight is thus more than merely the unproblematic functioning of a sense, and extends, like that of the 'solitary' man brought into society, into the gaining of culturally shared values. For both of these men, the sense of sight provides access to larger forms of social perception, and hence the new moral and aesthetic 'ways of being' that Smith describes for each.

In neither of these examples is the acquisition of sight (or moral insight) represented as in any way problematic, or as anything other than a straightforward learning process potentially open to all. For the newly socialised solitary man, a moral sensibility can be gained simply by learning the significance of the expressions and looks of his social peers; the value of his own social acts is visible in the 'mirror' of others' countenances, an image which suggests a near-automatic replication and circulation of social meaning. His tale represents a parable of the availability of a social currency of feeling, which, like a language, is open to being acquired by any social agent, and which is often learned simply by looking. As a compelling tale of the accessibility of social feeling, and an almost inevitable 'infection' by moral sympathies, it will be replayed again and again in sentimental literature, and extends even to the educative feelings described by the ultimate outcast, Frankenstein's creature, the 'sensations of a peculiar and overpowering nature', the 'mixture of pain and pleasure, such as I had never before experienced', on his viewing of the affecting tableau of the virtuous domesticity of the De Lacey family in Mary Shelley's *Frankenstein* (1818).[41] Despite the play on the dangerous harnessing of the powers of life-creation constituted by Frankenstein's scientific experiments earlier in the text, Frankenstein's creature, like Smith's solitary man, arguably only truly comes to life through this crucial experiencing of social passions: it is through his moral and social awakening that he becomes a truly vital being, as well as a potential recipient for readerly sympathy. For Cheselden's young patient, meanwhile, the acquisition of sight is both the restoration of vision and an inculcation into the cultural meaning and value ascribed to certain 'prospects', but whilst his viewing of such a 'prospect' might signal a privileged social and class status – given that the aesthetic appreciation of a prospect is associated with the leisured gentleman – there is no sense that such 'knowledge' is difficultly or problematically gained.[42]

The generation of sympathy – its circulation between and amongst a group of social agents to define and create their social interdependence and shared moral judgements – appears very different, however, in Wollstonecraft's *Short Residence*, where the moral animation gained through imaginative sight is often both more difficult for the narrator to secure during her foreign travels, and, given the work's failed revolutionary context and its quest for reasons for renewed hope, all the more valued. Travelling in regions which can appear so backward as almost to be out of time, the question of sympathy's necessary generation appears not – as in Smith – merely a matter of acquiring already-existing shared social values, but rather of finding some way to trigger its renewed growth, even in the face of the accumulated disadvantages of climate, environment and history.

The importance, as well as the difficulty, of generating sympathy is evident from the very beginning of Wollstonecraft's text, in an opening letter which replays, with a difference, Smith's (and Shelley's) bringing together of the solitary and the socialised, and which places the same value on the operation of mutual social understanding. Wollstonecraft's letter presents her arrival in a desolate landscape peopled by brutish and 'sluggish inhabitants'; as a solitary traveller who understands herself to be propelled from a more to a less developed society, the letter reverses Smith's depiction of the insertion of the uncivilised loner into more socially polished circles. Equally, the operation of an imaginatively informed sight, which in Smith guarantees a smooth communication between the parties, is in Wollstonecraft a much more fraught and elusive entity. The acquisition of forms of perceptive vision, which in both narratives for Smith was relatively problem-free, in Wollstonecraft becomes caught up in complex questions about generation. In these, imagination figures both as a product, capable of being created, replicated and generated, and hence an agent in historical transformation and human progression, and as exactly the opposite: as something always already there, beyond mere historical production, and therefore somehow without origin, and incapable of facile regeneration or easy deployment as an instrument or vehicle of historical change. The question of sympathy's generation, therefore, poses not only the issue of producing and maintaining the progressive forms of human feeling on which, in a post-revolutionary era, mankind's future hopes lie, but also raises the question of generation itself – of rebirth, renewal and progression into new forms of life – questions which are answered fittingly enough, if in rather a displaced way, in Wollstonecraft's meditations on motherhood later on in the text. Sympathy, it appears, lies problematically both within history and outside it: needing regeneration whilst already somehow present, with Wollstonecraft herself in the role of potential reanimator, the resuscitator of something feared as apparently dead, yet nevertheless always somehow living imperceptibly on.

The importance of vision, and the forms of perception and insight which it might enable or mediate, is evident as Wollstonecraft recounts her arrival in Sweden. Eagerly scanning the coast for possible landing points after 'contrary winds' blew her ship beyond more accessible ports, Wollstonecraft's sight is enthusiastically, but mistakenly, informed by an anticipatory imagination which repeatedly identifies possible harbours, only to have such visions disciplined by the empirical realities of what can in fact be seen. Searching for a 'boat to emancipate me', '[e]very cloud that flitted on the horizon was hailed as a liberator, till approaching nearer, like most of the prospects sketched by hope, it dissolved under the eye into disappointment'.[43] A looked-for liberation is not matched by given actualities; imagination's possibilities are failed by sight. But Wollstonecraft's anticipatory visions, for all the disappointments they bring, are immediately contrasted with the more limited sight of the natives she encounters when ashore, who fail to look about them even at what can very evidently be seen:

> Approaching a retreat where strangers, especially women, so seldom appeared, I wondered that curiosity did not bring the beings who inhabited it to the windows or door. I did not immediately recollect that men

who remain so near the brute creation, as only to exert themselves to find the food necessary to sustain life, have little or no imagination to call forth the curiosity necessary to fructify the faint glimmerings of mind which entitles them to rank as lords of the creation. – Had they either, they could not contentedly remain rooted in the clods they so indolently cultivate.[44]

The inhabitants' failings of sight are the failings of imagination and curiosity which characterise those living near 'brute creation', a depiction which only throws Wollstonecraft's own anticipatory forms of imaginative cognition into relief, and underlines her more 'developed' state. Wollstonecraft's organisation of human society into a hierarchised 'progression' from the 'brute' to the polished echoes the narratives of progression in Smith's essays on philosophical enquiry, or his tale of the solitary man, but his faith that the solitary will learn from the 'backward' is not shared by Wollstonecraft. Her experience of the failure of the Swedes even to look, let alone imagine, counters Smith's insistence on sight as a means of educative development. Their condemnation to the 'indolent' cultivation of clods amongst 'huge, dark rocks' which look like no more than 'the rude materials of creation' speaks for an impeded social and human progress, and sets forth the quandary of how such progress might be initiated, which will occupy the remainder of the book.

Imagination is of course in some way the answer here, and Wollstonecraft's language itself indicates its redemptive, generative possibilities. Her unusual word 'fructify', here used as a transitive verb suggesting a (masculine) active impregnation, rather than a (feminine) passive bearing fruit, gives imagination, and the curiosity it stimulates, an engendering role, but the lack of imagination amongst the population she describes turns attention from what imagination might generate, to what might generate imagination. What is figured here as a male principle of generation must itself be generated in order to kick start the necessary processes of curiosity and progress, but this must be achieved in a place where, Wollstonecraft comments, life itself seems somehow in abeyance. Wollstonecraft's own presence here, as an imaginative agent, might be thought to solve these questions of history, and certainly her restless activities recounted in the remainder of the letter – rambling in the near surroundings, going reluctantly to bed only to wake early with a still busy mind – could be understood to present the 'active principle' which is so clearly needed. Yet Wollstonecraft's imaginative activity remains notably enclosed and self-referential, failing to connect her to the 'sluggish' inhabitants whose lack of mental stimulus she has already diagnosed as their primary problem; instead, she prefers to contemplate an alternative passive object – the product of an alternative form of creation – rather closer to home. And just as it is her own realised creation, in her daughter, rather than the Swedes' failed self-cultivations, which she contemplates, it is also her own heightened animation, not theirs, which is produced by her ecstatic reveries:

> What, I exclaimed, is this active principle which keeps me still awake? – Why fly my thoughts abroad when every thing around me appears at home? My child was sleeping with equal calmness – innocent and sweet as the closing flowers. – Some recollections, attached to the idea of home, mingled with reflections respecting the state of society I had been contemplating that evening, made a tear drop on the rosy cheek I had just kissed; and emotions that trembled on the brink of extacy and agony gave a poignancy to my sensations, which made me feel more alive than usual.[45]

Here the foreign and abstracted – 'the state of society' – gives place to the domestic and personal, and an imaginative sensibility which, according to Wollstonecraft's earlier thoughts, might have revitalised the unanimated, contemplates instead what has already been brought to life. The problems contained by Wollstonecraft's thoughts on 'the state of society' are displaced by the alternative object of her daughter, through whom such questions about human progress will be replayed, in more personalised and potentially fraught terms, later in the text. If a set of questions about one form of generation has not been answered here, at least another kind of vital life has been presented in its place.

For all its evidence of her own active sympathies, and her female body's 'fructifying' powers, however, this reverie does not solve the underlying question of the origin of imaginative sensibility, the generation of the principle of generation. The 'recollections, attached to the idea of home' which bring tears to Wollstonecraft's eyes already refer, with their allusion to her daughter's father Gilbert Imlay, to absent origins, and this preoccupation comes to the surface as Wollstonecraft turns to the origins of sympathy itself. Remembering how often she has felt alienated from friends or the world, she questions the nature of the 'imperious sympathies', the 'involuntary sympathetic emotion' whose unknown and mysterious operation stops her seeming 'a particle broken off from the grand mass of mankind' and, 'like the attraction of adhesion, made me feel that I was still a part of a mighty whole, from which I could not sever myself'.[46] This assertion of an incontestable experience of the uncontrollable sympathies which return her to humanity – which reconnect her to the 'mass of mankind' – is also something which lies beyond human comprehension and even, as 'imperious' and 'involuntary', human will. The origins of sympathy, of the imaginative principle which will provide the 'fructifying' force to propel mankind beyond its present conundrums into a better future, lie beyond human control. This is the implication too of the letter's own, more immediate experience of involuntary human sympathies, which, by its conclusion, has entirely recast its initial condemnation of a 'brute' and undeveloped population into a celebration of the peasantry's 'golden age' simplicities, including its ready 'fellow-feeling' and 'overflowing of heart'.[47] Still presented as inhabiting a different era in the development of human society from her own, the Swedes no longer occupy a place disturbingly close to 'brute creation' but a mythic 'golden age' beyond history, and the problem of human history – of human imagination and sympathy – is dissolved in the same turn to myth. The existence, after all, of golden age sympathies and fellow-feeling removes the problems of the necessary 'fructifications' of imagination as the key to human progression, as such impulses and sentiments are already present in humanity's mythic prehistory. The 'golden age' fellow-feeling of the peasants, like the involuntary imperialism of sympathy noted later, counters Wollstonecraft's attempts to provide a historical analysis of (and solution to) imagination's necessary generation, and, if only briefly, soothes her fraught sense of an impeded human progress: the 'mythic' presence of sympathy, and hence imagination, asserts that origins need not be sought, that they are always already present, and thus that, whilst their generation fuels the progress of history, they also lie beyond and outside history too.

It is not only in a myth which is both of and outside human history that the question of generation finds an answer: the materiality in which human life consists also points to some solutions on this topic. Wollstonecraft's expression of herself as a 'particle', newly 'adhering' through involuntary sympathy to the 'mass' of mankind, figures sympathy decisively as a material force – a metaphor which is all the more surprising given that her meditations elsewhere on the unknown forces of energy and life, whilst they use materialist language, seem to do so only reluctantly. Musing on the passing of human life in response to seeing the preserved bodies in Tønsberg, she exclaims, 'Life, what art thou? Where goes this breath? This *I*, so much alive? In what element will it mix, giving or receiving fresh energy? – What will break the enchantment of animation?' An attempt to articulate the mysterious something, the more than material, which animated the dead bodies she now sees, falls back on a language – 'breath', 'element', 'energy' – which is never fully metaphysical. Part of the 'enchantment of animation', it might appear, is the way in which the mysterious force of life – which might also be described as, like that of sympathy, 'imperious' and 'involuntary' – cannot satisfactorily be expressed as an entirely metaphysical entity: 'enchantment', the nearest Wollstonecraft gets to this, is a merely magical, 'occult' term. The words which communicate this uneasy verbal residue of the material, meanwhile, are spoken at a scene which itself presents disturbing evidence of the physical residue of past human lives. Acting against the characteristic movement of her travel writing, which has everywhere sought to discover, expose, explore and penetrate, Wollstonecraft declaims in Burkean fashion against the treasonable lifting of 'the awful veil' which would 'fain hide [humanity's] weakness': the disturbingly persistent residue of mankind's weak materiality represented in the preserved bodies is, it seems, best draped and repressed.[48]

Other forms of unwanted human traces, of a rather different kind, addressed elsewhere in the text also prompt an unwelcoming response from Wollstonecraft – even whilst returning her again to the theme of sympathy and its material consequences. Her thoughts on this occasion are prompted by observations of the prevalence of young unmarried mothers in Sweden. '[C]asual sympathies of the moment', Wollstonecraft suggests, might be what has produced the material evidence of these women's 'attraction', if not to the 'grand mass' of mankind, at least to a part of it; a rather less welcome realisation of the 'involuntary' and 'imperious' operation of sympathy than she celebrates elsewhere, and one whose physical residues, to put it bluntly, provide the same evidence as the bodies at Tønsberg of what for Wollstonecraft appears to be human weakness and frailty. If sympathy's generation had earlier been suggested as a solution for Swedish 'sluggishness', its operation between human bodies, rather than human minds, is clearly antipathetic for Wollstonecraft in such cases. Yet Wollstonecraft's addressing of her own daughter, herself the offspring of an unmarried alliance, celebrates both her physical, material qualities and a projected imaginative sensibility, and suggests that the difference between the basely material and the imaginatively elevated is less stable than she would seem to want. For all her concern over the progress of human society, which for her is ultimately dependent on the proper 'fructifying' activity of the imagination, her own daughter, and perhaps those of the unmarried Swedish women addressed here, represents an alternative solution to the problem of generating progress which so occupies Wollstonecraft in this text. In the apparent absence of a grand, masculine, imaginative 'fructifying' principle, through which a 'current of life' which 'seemed congealed at source' will be brought to 'enchanted' life, the more 'casual' sympathies of Swedish women, with their material by-products, might appear to represent their own generative possibilities.[49]

For Wollstonecraft it would seem that, whilst on the one hand imagination presents problems of impossible generation, on the other, the action of inevitable, 'involuntary' sympathies already present their own answer to such questions of origin, progress and generation. Where the maternal body, in *Vindication* and *Mary*, is weak and degenerate, what could be seen in the context of those texts as its greatest weaknesses – its involuntary motions, its failure to be rationally controlled by the conscious will, its 'sensibility' – could be seen, in the context of the problems and quandaries of *A Short Residence*, to be its greatest strength. Where the *Vindication* attempts to discipline the body, and plots the emancipation of its subjects through more mental means, in *A Short Residence* the body's 'involuntary' sympathies present an alternative solution to the problem of human progress which the 'active principle' of the imagination cannot always resolve. At a time when the mind is baffled and barren, yearning and restless, despite all its activity, the body provides an alternative, material, involuntary principle of sympathy and animation, of vitality and progress, capable of redeeming and renewing human history even in the face of seemingly impossible hopes. The unsuppressible vitality of human bodies both makes evident the existence of past sympathies, and figures forth the possibility of future ones.

3
Labouring Bodies in Political Economy: Vitalist Physiology and the Body Politic

Given in full, the usually abbreviated title of Adam Smith's *An Inquiry into the Nature and Causes of the Wealth of Nations* (1776) outlines with the clarity characteristic of its author the work's precisely defined concerns. Identified with a simplicity which the work's near-thousand pages belies, *Wealth of Nations* answers the question its title poses in its very first sentence, but in such a way as to suggest the need for the two volumes of further analysis which await the reader. On his opening page, Smith defines the nature of a nation's wealth as 'all the necessaries and conveniences of life which it annually consumes' and the origin of that wealth as '[t]he annual labour of every nation'.[1] Stated thus, Smith's perception that the labour of a nation is also its wealth is plain, but the complexity of what labour might be, the different forms of activity which might constitute it, the relation between the labour of an individual and the wealth of a nation, and so on, becomes evident as the introduction progresses. If Smith's original statement, that a nation's wealth consists in its labour, might seem foreclosed, once lines of discussion and analysis are sketched, and the bewilderingly varied range of activities which might be described as forms of labour are described, the term 'labour' emerges as an abstraction only possible through an almost wilful forgetting of the variety of material and intellectual practices which it might name. Labour, in this sense, is a collective noun whose analytical value lies precisely in its ability to mediate between, on the one hand, the specificity and differences of labouring acts, and the requisite abstractions of Smith's economic analyses on the other. It is a term which translates the historical and varied reality of acts of toil, labour and effort into the abstracted form of the significance of those acts within the terms of the political economy which Smith is here writing.

The different ways in which the term 'labour' operates in *Wealth of Nations* can be demonstrated within its first hundred or so pages. Smith's first chapter, on the division of labour, offers a provoking and accessible illustration of the variety of different kinds of labour in a 'civilized and thriving country' through its meditation on the rough and coarse 'woollen coat' of the day-labourer. This, as Smith explains, is 'the produce of the joint labour of a great multitude of workmen':

> The shepherd, the sorter of the wool, the wool-comber or carder, the dyer, the scribbler, the spinner, the weaver, the fuller, the dresser, with many others, must all join their different arts in order to complete even this homely production.[2]

To this, Smith then adds the 'merchants and carriers' who transported materials, and the 'ship-builders, sailors, sail-makers, rope-makers' who have imported the drugs used by the wool dyers, before adding to the increasingly lengthy list those employed in making the tools used in making the coat:

> What a variety of labour too is necessary in order to produce the tools of the meanest of those workmen! To say nothing of such complicated machines as the ship of the sailor, the mill of the fuller, or even the loom of the weaver, let us consider only what a variety of labour is requisite in order to form that very simple machine, the shears with which the shepherd clips the wool. The miner, the builder of the furnace for smelting the ore, the feller of the timber, the burner of the charcoal to be made use of in the smelting-house, the brick-maker, the brick-layer, the workmen who attend the furnace, the mill-wright, the forger, the smith, must all of them join their different arts in order to produce them.[3]

In this passage, Smith's writing is exhortative and descriptive, a catalogue rather than an analysis. Where meditation by Shaftesbury might enthuse over the order and beauty of the appearances of nature, and their manifestation of divine wisdom, the wondrous appearances here are of human labour animated – in this instance – by the labourer's humble need for a coat. The survey of the variety of labour, as opposed to nature, manifests not a divine order, but a complex and intersecting network of human activity, whose 'order' might in fact be understood in a variety of ways: not simply as the production of a workman's coat, but as the full range of manufacturing, trading and commercial activity of a 'thriving' country, in which Smith might just as easily have traced the discrete but connected labour involved in the manufacture of a shoe-buckle, or the production of – a favourite example of his – a watch. The 'order' manifested in the nature Smith sketches here, in fact, is that of a system of human labour organised to meet a bewildering number of human needs and desires. Meanwhile, although this passage functions to illustrate 'labour', it also demonstrates what Smith repeatedly reminds us is the sheer 'variety' of activity which might be included under that term, a variety which might threaten not only the coherence and organisation of his own writing in this passage, as it descends into an amazed listing, but also the equivalences between different forms of labour which *Wealth of Nations* elsewhere seeks to calculate and establish. As Smith's examination of the kinds and types of labour proceeds, the unity of the term 'labour' itself comes under pressure from the range of practices and acts which he himself shows us it might name.

Here at the outset of *Wealth*, 'labour' as a term is closely connected to human activity. It can also ascribe identity, as some forms of labour are evoked simply as the 'arts' of individuals who are named by referring to the forms of work they do: the shepherd, the scribbler, the dyer and so on. Elsewhere, Smith also recognises the frequently bodily nature of labour at this time too: labour is the 'toil of our body', and the wages of labour often reflect its ease or hardship, cleanliness or dirtiness.[4] The cost of labour in terms of the long-term health of the body is also noted, with a reference to the Italian physician Ramazzini's work, *A Treatise of the Diseases of Tradesmen*.[5] As Smith's text gets underway, however, the material variety and bodily effort of labour is increasingly noted in order to calculate their significance in economic terms. Smith's discussion of the wages of labour, in chapters eight and ten, for example, retains the commitment to a detailed recognition of the material reality of the life of a labourer which characterises his opening chapter, but here his concern with the feeding, the clothing, even the reproduction of the workforce, is part of a primarily economic calculation, necessary to determine the fairness of current wages. Where, in the opening chapter, the sheer variety of labour carried out in a modern commercial nation is Smith's primary theme, here variety is noted only to give way to calculations which efface the material differences of labour by translating them into forms of economic equivalence. Thus, Smith compares the easiness of the work of a tailor and a weaver, the relative dirtiness and danger in the work of a blacksmith and a collier, and the 'lottery', both in terms of losing one's life and gaining glory, for a soldier and a sailor, not, as before, to marvel at the variety of forms of work in a thriving nation, but to suggest that these forms of difference can be rendered in economic terms through the varying rates of recompense given for each kind of work.[6] This development in Smith's discussion shifts the significance of the term 'labour': whilst retaining a recognition of it as a human, and bodily toil, 'labour' becomes a term in a discourse which, by calculating an economic value in labour different from, but equivalent to, its material reality, translates the human variety and difference of work into a calculable and abstracted economic value. The effect of this development in Smith's discourse, away from the wondrous cataloguing of human effort, to the translation of that effort and activity into a purely economic significance, can be seen even at the level of his language, which expresses basic human realities, of subsistence, survival or habitation, in sometimes startling terms: it becomes possible, for example, to speak of the 'wear and tear' of a slave or servant, or to note that, given a reluctance amongst workers to move from one part of the country to another in search of better wages, 'a man is of all sorts of luggage the most difficult to be transported'.[7] In these examples, as Smith's language betrays, the human has become object, as the experience of the human subject lying within each scenario is effaced by Smith's translation of that experience into its abstracted economic effect.

This shift in the usage of the term 'labour', from a name which collects a range of specific acts of human toil, to an abstraction amenable to economic analysis, of course expresses the nature of the kind of text which Smith is here writing. Proclaimed on its publication as the first complete and systematic account of national economic activity, which had previously been subject to more piecemeal analysis, Smith presents his work as a contribution to the science of 'political œconomy', the study of which became instituted, a few years after his death, at Edinburgh University with the lectures of Dugald Stewart. Whilst *Wealth of Nations* often concurs with accounts of aspects of economic activity offered by earlier commentators – Smith's identification of labour as the origin of wealth, for instance, is shared by Hume, and can be traced back to Locke – its originality lies in its presentation of the economy as a connected system, in which the links between labour, wages, the circulation of money, the value of land, the effects of manufacture and trade, and the role of government can all be traced. In part, such links are demonstrated by precisely the abstraction from material differences to economic effect or value which we have noted in Smith's treatment of labour: the analytical power of Smith's new political economy lies partially in demonstrating how equivalences and exchanges between material differences (different kinds of labour, different kinds of commodities) might be calculated, and how the economy constitutes just such a system of mediated exchange between differences. The nature of commerce, after all, is the exchange of commodities (or embodied labour, as Smith at times regards them), via a system of calculated equivalent values, expressed in monetary terms. The treatment of labour which has been traced here, then, where Smith balances an acknowledgement of the material reality of labour – variously, its dirtiness, danger and differences – against a calculation of its significance in economic terms, as abstract value, is characteristic of the analysis offered by *Wealth of Nations* as a whole, whose own analytical value lies in moving beyond the disparate appearances and experiences of the world to discover a hidden system of connections uniting them into a cohesive 'œconomy'. Such a movement of uncovering hidden connections is what Smith himself understood as the particular job or 'trade' of the philosopher: whilst, in the age of the division of labour, other workers were confined by the demands of efficiency to the repeated performance of the same few specialised actions, the job of philosophers or 'men of speculation' was 'not to do any thing, but to observe every thing', and they therefore, 'upon that account, are often capable of combining together the powers of the most distant and dissimilar objects'.[8] The context of Smith's remarks here makes it clear that he is considering both philosophers like himself and the inventors of the kinds of 'ingenious' machines which were beginning to transform manufacture in the second half of the eighteenth century, both of whom produce useful and beautiful 'systems'. Smith's observation thus offers its own unexpected connection, between two forms of labour which might appear very different, as well as reinforcing Smith's sense, also expressed in his early essay on the nature of philosophical enquiry (discussed in the previous chapter), that the systems created by philosophers, like those of inventors, are machines which find out and link the unknown or hidden forms of connection between differences, to powerful, even transformative effect.[9]

In its analysis of labour, *Wealth of Nations* describes a systematic means of mediating between the material and the abstract.[10] But even as an abstraction, labour at some level retains a memory of its material origins in human effort; thus to tell a story about labour, as *Wealth of Nations* does, is also to tell a story about the human subject, who, like labour itself, is also constructed, produced and represented by the discourse of political economy, both as a material, bodily subject, and as the subject of a more abstract knowledge of human nature. Political economy's interest in labour thus demands a representation of the human subject: labouring bodies imply persons who labour, and whose need, desire and motivation to labour must be understood and accounted for. Further, political economy's account of the labouring subject must be linked to the consequences of that labour for the wealth of the nation; just as, in its account of labour, it abstracts from material effort to economic value, the particular labour of the subject must be linked to the generalised economic consequences of the cumulative labour of subjects. To recognise the importance of representations of the human subject in political economy is not new: our understanding of competing arguments for and against the new commerce in the early eighteenth century has been enriched by J. G. A. Pocock's demonstration of the opposing models of human personality which were deployed in such debates; Albert O. Hirschman too has traced ways in which economic thinkers from the Renaissance onwards considered ways in which human passions might be 'harnessed', exploited or controlled in commercial activity.[11] The particular concern of this chapter, however, is to trace ways in which the theorisation and writing of Smith's political economy itself depends on the models of the human subject which it constructs and deploys, in part by adapting accounts of man's passions, psychology and physical nature available in other philosophical discourses, including, most importantly, vitalist physiology. The 'abstract' and general knowledge of human nature necessary to political economy is thus at times derived from much more material accounts of what it is to be human.

Wealth of Nations' representation of the human subject is often offered through an account of the human body, and not only because it is most typically concerned with the subject whose labour is bodily. Its representation of the economic subject is frequently of a subject embodied: a subject whose passions, motivations and desires might be known and expressed via the body and have particular bodily consequences and effects. This means that to address ways in which Smith and others sought to represent the subject within economic discourse, as this chapter does, is often to be thrown back onto accounts of the body, health and disease, or to be caught up in terms which, like 'pleasure' and 'indolence', do not refer exclusively to the body, but suggest an experience which might in part be a bodily one. To examine the placement, figuration and representation of the body, and hence the subject, in *Wealth of Nations*, also presents itself as a way of throwing into relief the particular representational manoeuvres of political economy in a formative, foundational moment: to consider not only its depictions of the subject and subjectivity, and the relationship between the subject and the collective whole of the nation or economy, but also the constitutive metaphors of political economy itself, and its relation to other discourses whose representations of the body it often borrows and reworks. It is to consider, too, the modes by which the human subject and human nature are represented within a discourse to which such concerns are necessary, but which might also at times appear to efface or overlayer them with larger concerns of political economy. Representations of the subject and the body might thus be considered as a faultline running through *Wealth of Nations*, a means of accessing the constitutive narrative or representational strategies of political economy itself. The next section addresses ways in which accounts of the subject in economic writings by Smith and others are produced via various different models of the body; the second half of the chapter explores ways in which the economy itself is understood by Smith as like a very particular kind of body, the 'vitalist' body currently being theorised by Smith's contemporaries at the Edinburgh Medical School in the mid-eighteenth century. The chapter thus shows not only how the human subject is written in economic discourse, in part through representations of the body, but also how Smith's defining narrative of political economy constructs an economy which is both the product of labouring bodies, and is itself a very particular kind of body whose 'health' or well-being, like that of the body in vitalist physiology, is best ensured by leaving it to its own self-regulation.

Labouring bodies in economic discourse

Smith's identification of labour as the foundation of commercial prosperity is not of course unique to him, but rather might be placed in a tradition of philosophical discussion of labour which could be traced back to Locke's assertion, in his essay 'Of Property' in his *Second Treatise of Government* (1690), that labour is the origin of economic value. As Tom Furniss has demonstrated, Locke's recognition of the economic significance of labour, however, only highlighted a further set of questions: what is it which motivates man to labour, what is the moral status of such motivations and how might such motivations be controlled?[12] The recognition of the economic importance of labour was thus connected to a set of questions about human passions, desires and psychology, and a model of human subjectivity which accounted for the 'springs of action' became necessary. Locke himself, in the first three editions of his *Essay on Human Understanding* (1690), suggested that recognition of 'the good' which results from it is what spurs man to work, but he replaced this optimistic account with a rather darker one in his fourth edition. '[U]pon a stricter enquiry, I am forced to conclude, that *good*, the *greater good*, though apprehended and acknowledged to be so, does not determine the *will*, until our desire, raised proportionably to it, makes us *uneasy* in the want of it.'[13] For Locke, uneasy desire locked man into a constant willingness to work, as the means of best satisfying his wants, of which new ones formed just as others were met. The only route out of the vicious circle was the cultivation of discrimination and taste, which enables the suspension of desire, but such a solution could only be available to the leisured, paradoxically those who perhaps least needed it.[14] For a later writer, Mandeville, meanwhile, it is a specific kind of desire which motivates work: it is 'Man's natural Love of Ease and Idleness' which is the foundation of his labour.[15] If Locke sees man as caught in an uneasy and endless cycle of desire and labour, Mandeville places man in the contradictory position of labouring only so he may labour no more.

A particularly detailed consideration of the 'springs' of labour is given in David Hume's economic essays, in an analysis which draws also on his account of the passions in Book II of his *Treatise of Human Nature* (1739).[16] Hume appears to have shared Mandeville's perception of mankind as naturally indolent and ease-loving, but an ingenious argument is used to demonstrate how, in effect, ease itself transforms an unpromising populace into a motivated and industrious workforce. Ease itself, it is suggested, follows a cyclical pattern, so that in time it becomes tedious and 'uneasy', demanding action and exertion before relaxation and indolence can again be enjoyed. Cleverly, Hume dispenses with the need for any, potentially problematic, external stimulus to prompt labour: it is human nature itself which needs work, and whatever the beneficial economic by-product of that work, ultimately it is man's own need for exercise and exertion which labour serves.

Hume's account of ease depends on his theory of the passions, articulated at greatest length in his *Treatise of Human Nature*. Here, passions are understood in terms of cycles from one passion to another, best demonstrated in the description of the transition from novelty to tedium. To undertake an action, or experience an unknown object, for the first time, involves 'difficulty': pain, exertion and uneasy passions. If the action or experience is repeated many times, however, these passions are converted into the more pleasurable ones of ease and familiarity, as we become accustomed to whatever it is which has prompted them. The cycle does not stop here, however, for, 'as facility converts pain into pleasure, so it often converts pleasure into pain', and objects and actions which are experienced too frequently can become tedious and tiresome 'thro' custom' itself. Thus, as unease can become ease through custom, so, in the same way, ease can become unease, for 'nothing has a greater effect both to encrease and diminish our passions, to convert pleasure into pain, and pain into pleasure, than custom and repetition'.[17] Hence, the ease which for Hume is fundamental to human nature clearly follows a cyclical pattern, so that action, as well as rest, may be experienced as pleasurable. According to the *Treatise*, man is caught up not in a cycle of desires prompted by external objects, but in a cycle of passions which constitute human nature itself. This perception underlies Hume's definition of human happiness in his essay *Of Refinement in the Arts*, where it is said to consist of 'three ingredients; action, pleasure, and indolence': 'no one ingredient', he says, 'can be entirely wanting, without destroying, in some measure, the relish of the whole composition'.[18]

Although this cycle of successive desires might appear to favour action and inaction equally, in fact Hume suggests that it tends increasingly to encourage action. Familiarity, he suggests, gives 'an inclination and tendency' towards the performance of a particular action, whilst, at the same time, familiarity with inactivity decreases the desire for it: custom 'encreases all *active* habits, but diminishes *passive*'. The cycle of activity and inactivity is thus increasingly skewed towards the performance of labour, and away from rest and indolence – whilst exertion is still experienced with pleasure. In this way, a human nature governed by desire for ease may be seen to be at work even in modern commercial society.[19] Through custom, 'industry' becomes increasingly pleasurable, and increasingly pursued in preference to indolence. A 'reward' in itself, the labour of commercial society is, according to Hume, not a diversion from pleasure but a pleasure and form of fulfilment in itself. Through it, '[t]he mind acquires new vigour; enlarges its powers and faculties; and by an assiduity in honest industry ... satisfies its natural appetites, and prevents the growth of unnatural ones'.[20] In commercial society, man has not only risen above his natural state of indolence, but he undertakes the activity necessary to the prosperity of that society out of preference, as a form of pleasure.[21] In this way, the historical change from pre-commercial to commercial society is understood as a generalised consequence of the cycle of indolence and activity, with its progressive and increasing encouragement of action. A naturally indolent mankind increasingly finds pleasure in the exertions and efforts of industry, and hence is transformed into the productive workforce which Hume's economic theory, which like Smith's is founded on labour as a source of prosperity, finds so necessary.

The language used by Hume here, which describes the motions of the passions as producing the energy or force of motivation, able to exert itself on the passivity of the will, is that of a mechanical model, in which change only occurs through the application of a force strong enough to overcome the resistance of inertia. Among many commentators in the debate over Hume's Newtonianism, Christine Battersby has suggested that Hume's account of the mind, which offers what she characterises as an 'epistemology of ease', is based on Newtonian principles of inertia.[22] Battersby does not explicitly address Hume's treatment of the passions, but the mechanistic model which could be seen to inform his discussion of them in Book II of the *Treatise* becomes evident in a later recasting of its material. In his dissertation 'Of the Passions', published in 1757, Hume argues that the passions are subject to the same 'certain regular mechanism' as is described in the 'laws of motion, optics, hydrostatics, or any part of natural philosophy'.[23] Certainly, Hume's account of the motions and forces of the passions, in which the static state of indolence produces in itself the motivation to labour, without the application of any external force, might recall the discipline of hydrostatics, a key element of Hume's natural philosophy studies at Edinburgh University, which examined the equilibrium of liquids and the pressure exerted by them at rest.[24] But regardless of such specific analogies and possible sources, Hume's account of labour clearly constructs a subject for whom work is less an object through which other desires might be satisfied – a necessary, inevitable and unwelcome undertaking through which a whole nexus of subjective desires might be believed to be met – than one whose emotional machinery dispassionately converts indolence to labour, a conversion which recalls the eighteenth-century understanding of a machine as a device for transforming one kind of motion or force into another. Hume's account of labour, in short, sees the individual less in terms of the complex and interlinked desires, motivations and consciousness which might constitute a subjectivity, and more in terms of a machine, in this eighteenth-century sense, whose indolence almost inevitably converts itself to labour, regardless of any intervening desires or determinations of the subject.

Smith's account of the labouring subject shares some aspects with Hume's but is distinct too in other important ways. If Hume offers a mechanical account of motivation, in which opposing forces of ease and unease generate an increasing momentum towards a 'pleasurable' work, Smith by contrast depicts a labouring subject in whom similar patterns of effort and relaxation are particularly seen as serving a physical purpose: as meeting the body's 'natural' need for work. Rather than the exertions of work being prompted by the pressures of the passions, as for Hume, labour in Smith is considered in relation to the natural needs and health of the body. This analysis is clearest in Smith's discussion of labour wages, in which he argues, in contrast to some contemporary commentators, that high wages will be a source of encouragement and motivation for the workforce. Whilst in some cases, he admits, 'some workmen ... when they can earn in four days what will maintain them through the week, will be idle the other three', in general, 'where wages are high ... we shall always find the workmen more active [and] diligent' than where they are low.[25] Better-paid workmen are better-fed workmen, their 'bodily strength' increased by 'plentiful subsistence', but, although high wages thus suggest a physically robust workforce, there is another danger: so encouraged are they by liberal wages that 'the greater part' of workmen are liable to work excessively, even to the point of endangering their health. A carpenter in London, for example, might burn himself out after a mere eight years of labour. High wages promote labour, then, but might also cause the two extremes of under- and overwork, both equally unproductive. Caught between the dangers of encouraging idleness on the one hand and overwork on the other, Smith turns to the body to secure his argument, and suggests that its own natural demands for work and rest, if allowed to govern labour patterns, would both safeguard the worker's health and ensure maximum productivity:

> Great labour, either of mind or body, continued for several days together, is in most men naturally followed by a great desire of relaxation, which, if not restrained by force or by some strong necessity, is almost irresistible. It is the call of nature, which requires to be relieved by some indulgence, sometimes of ease only, but sometimes too of dissipation and diversion. If it is not complied with, the consequences are often dangerous, and sometimes fatal, and such as almost always, sooner or later, bring on the peculiar infirmity of the trade. If masters would always listen to the dictates of reason and humanity, they have frequently occasion rather to moderate, than to animate the application of many of their workmen. It will be found, I believe, in every sort of trade, that the man who works so moderately, as to be able to work constantly, not only preserves his health the longest, but, in the course of the year, executes the greatest quantity of work.[26]

The body here operates as a natural organism whose predictable patterns will control and stabilise not only the labour force, but also the unruly elements of Smith's own argument. By suggesting that the body has its own natural demands for work and rest which should be allowed to regulate the individual's labour patterns, Smith's initial positing of high wages as the primary factor motivating the workforce is both supported and supplanted. The separate concerns of health and work become intertwined, so that the preservation of the one depends on the other: to work is to maintain one's health, and to maintain one's health is to work. By constructing a body with successive, regular and natural demands for both work and rest, Smith depicts a worker whose need for work is understood in terms of health rather than economic needs – his 'weal' rather than his 'wealth' motivates him – and Smith's initial external incentive of high wages becomes an irrelevance. The motivation to work becomes a natural need, the imperative to labour a 'call of nature', a simple bodily demand. Naturalised and internalised, the patterns of productivity necessary to the modern commercial nation appear to coincide neatly with providential designs for man, manifest in the innate working patterns of the body.

On this occasion, Smith's turn to the body provides a way of closing a wayward argument which had threatened to arrive at a conclusion other than that intended by him. These references to the body's internal demands and desires are later elaborated, however, as Smith supplements the initial picture of the worker called by nature to divide his time between work and rest, to describe him as also subject to a constant desire to better himself, an aspirational impulse which accompanies him from birth to death. No man, says Smith, is ever 'so perfectly and completely satisfied with his situation, as to be without any wish of alteration or improvement, of any kind'; rather, a 'comfortable hope of bettering his condition, and of ending his days perhaps in ease and plenty, animates [the worker] to exert his strength to the utmost'.[27] This motivation for self-betterment is presented as a constant ruling principle exerting its force throughout man's life:

> [T]he desire of bettering our condition [is] a desire which, though generally calm and dispassionate, comes with us from the womb, and never leaves us till we go into the grave. In the whole interval which separates those two moments, there is scarce perhaps a single instant in which any man is so perfectly and completely satisfied with his situation, as to be without any wish of alteration or improvement, of any kind.[28]

Where, for Hume, a cyclical, even mechanical operation of the passions fuels the worker's labour, Smith offers an alternative motivational force: a single, constant, aspirational energy, a naturalised, internalised, unrelenting drive which, as a constant presence between birth and death, might almost be seen as a defining characteristic of life itself. Self-interested, 'natural' and constant, depicting man as struggling unremittingly for self-preservation and betterment, Smith's account of the energies animating the labouring body is rooted in the same language of bodily forces and natural needs which earlier secures his defence of high wages. Although, like Hume, for whom the near-mechanical pressures of the passions result in the accumulation of a more or less constant desire for labour, Smith's model of the labouring body is one in which the need and capacity for work are certain, that model is founded in a very different discourse: of natural forces constantly animating man's unremitting desire and struggle for improvement. If Hume's model borrows from the mechanical systems of contemporary natural philosophy, Smith's depiction of man allies him to a world of animated beings defined by a single, lifelong and life-defining force or energy.

In these passages, Hume and Smith are constructing models of the human subject through which the origins of economically necessary labour can be understood, but the nature of their own labour, the work of the philosopher, also comes under consideration. In his discussion of the motivating causes of labour, Locke differentiates between those who are caught up in the cycle of desire which prompts them to work, and those philosophers, or men of leisure, who are capable of cultivating the rational self-improvements of 'taste' which release them from that cycle. As Furniss comments, the irony is that those most able to shake themselves free from the compulsion to work are those for whom the economic necessity to work is already least pressing.[29] Locke here allies the philosopher to the gentleman of leisure, suggesting philosophy is itself less labour than a form of self-improvement. For Smith, by contrast, philosophy is still a 'trade' or 'profession', and hence a form of work, but, although he contends that the differences between a philosopher and a street-porter are merely the consequences of education and upbringing, he nevertheless differentiates in significant ways between the labour of the philosopher and that of other trades. The labour of the philosopher, according to Smith, is to produce connected, explanatory systems of thought, and is driven, like that of more humble workers, by a desire for ease, but ease in this context is not a form of economic security or the happiness of consumption, but the ease of the imagination in aesthetic pleasure. Philosophy, according to Smith, is an art of the imagination: the imagination creates philosophical systems to aid its own 'smooth' and 'pleasurable' contemplation of the world. A connected philosophical system gives the imaginative pleasures of 'ease' and 'smoothness' as the philosopher contemplates the world; the philosopher's labour, in creating such systems, is thus, unlike that of other trades or professions, directly productive of his own pleasure, and by involving the philosopher in general views of broadly connected systems, rather than isolating him in the divisions of labour to which others are subject, he also escapes the alienating effects of work documented by Smith elsewhere. As the practitioner of a 'trade' or 'profession', the philosopher's labour is presumably understood by Smith to be motivated by the same desire for self-betterment and ease to which all are subject, but unlike other workers, for whom a desire for ease traps them in an endless cycle of work, the philosopher's work does provide a form of pleasure and release, albeit in the immaterial realms of the imagination.

Smith's discussion of the work of the philosopher redeploys the same language of effort and ease which informs his account of all economic subjects, but finds for the philosopher a form of fulfilment through that effort which eludes other workers. For Hume too, his own labour of philosophical writing repeats the same cycle of indolence, discomfort, effort and relaxation which he describes in the motivation of labour in general, but with the difference that, in his own case, there is no happy accumulation of effort into increasingly pleasurable production, but rather repeated mental exertions intercut with both indolence and despair. In both the *Treatise* itself, and Hume's autobiographical accounts of his work on it, the harmonious synthesis of indolence, effort and pleasure fractures into an account of human nature torn between disciplined effort and the wayward desire of body and mind for 'ease'. Whilst his account of the motivational passions of the labourer transforms the human love of ease into a productive cycle, philosophical labour is characterised in the *Treatise* as a struggle against the various natural determinations or propensities of the human mind: a rigorously sceptical philosophy must work against a principle of 'ease' by which the mind would prefer to believe something quite different from what philosophy demonstrates. Ease in the context of this kind of labour, in fact, does not stimulate production, as it does so ingeniously in Hume's later economic writings, but rather threatens to swamp philosophical production with 'delusion', 'weakness' and 'bias'. Such a struggle between philosophy and human nature is evident, for instance, in Hume's account of the conflict between our discrete, interrupted, sense impressions, and our contradictory (but philosophically false) notions of the continued existence of external objects. Our attempts to reconcile sense impressions with ideas becomes a paradigmatic instance of the struggle of the *Treatise* as a whole, as a rift opens within the subject himself, who experiences a 'sonic stuggle' between reason and philosophy on the one hand, and imagination and nature on the other. Attempting to resolve this rift, and 'set ourselves at ease', we produce the theory of a 'double existence of perceptions and objects', but this 'new fiction' is simply a 'monstrous offspring' of two contrary principles, reason and imagination, 'both embraced by the mind, and unable mutually to destroy each other'.[30] In an image which anticipates the monstrous product of philosophical labour in Mary Shelley's *Frankensten*, Hume's labour gives birth to an internally divided, self-conflicted human subject, whose attempt to resolve such differences, in pursuit of ease, leads only to the creation of monstrosity.

A similar sense of the unproductive efforts of philosophical labour recurs in Hume's more personal accounts of his experiences as a writer of philosophy. In both an early letter to a doctor, written during his composition of the *Treatise*, and the self-reflective last chapter of the work's first book, the same struggle of human nature with itself is evident, a struggle between philosophical endeavour and a desire for ease, between pursuit of reason and the 'illusions of the imagination' for which his nature 'conspires' to make him feel a 'strong propensity'.[31] Hume's letter of 1734, written some five years into his work on what was to be the *Treatise*, and narrating in detail the effects and symptoms of his sustained philosophical efforts, recounts a familiar cycle of a 'strong Inclination' to study and an 'Ardor' of spirit giving way over time to a 'Laziness of Temper' which must be 'overcome by redoubling my Application'. Understanding himself as suffering from 'the Disease of the Learned', Hume takes 'some Indulgence to myself', moderating his study with exercise and relaxation, and finds it possible to continue his labours, but faced with 'many a Quire of Paper, in which there is nothing contain'd but my own Inventions', and the need to organise these 'Parts in Order', his spirits again fail him. The letter concludes with his decision to turn from philosophy to the active life of a merchant, 'to toss about the World, from the one Pole to the other, till I leave this Distemper behind me', and can thus return to his studies.[32] Although rendered as a pathology, the same repeated cycle of effort, indolence and effort which informs Hume's account of others' labour is evident here, with the important difference that, for the philosopher, that cycle has been of effort and suffering without, at the time Hume writes, any increasing desire for labour, or any achieved production from it. The last chapter of Book I of the *Treatise* repeats a similarly unhappy narrative of Hume's philosophical studies, but recasts what was presented in terms of the bodily 'Disease of the Learned' into a psychopathology of melancholy and despair. Philosophical effort has 'heated' Hume's brain to a state of confusion and disorientation, in which he 'fancies' himself 'in the most deplorable condition imaginable, inviron'd with the deepest darkness, and utterly depriv'd of the use of every member and faculty'. The 'indolence' prescribed by nature offers some partial cure, but Hume nevertheless feels that he must 'strive against the current of nature, which leads me to indolence and pleasure', to return again to philosophical pursuits, even if such a determination is experienced as deformity and alienation, as he sees himself as 'a strange uncouth monster ... who not being able to mingle and unite in society has been expell'd'.[33] As in the letter, ease here is not part of the productive cycle of labour but a diversion from it, and in neither text is the struggle between ease and philosophy, nature and knowledge, happily or productively resolved: any philosophical output, other than such as these narratives of struggle themselves constitute, never appears assured. In both accounts, the dynamic is not of production but oscillation between effort and self-defeat. Whilst the philosopher reports in himself the same passions, desires and motivations which he uncovers in a more general account of human nature, the nature of his own work is far from a straightforwardly happy production.

In their description of the work of other economic subjects, Hume and Smith both construct models which offer generalised and naturalised accounts of man's need for and motivation to labour, seen either as a function of the mechanism of the passions, or as a natural desire, a life-defining urge characteristic of human nature itself. When applied to their own labour as philosophers, however, the same language, of cycles of effort and exertion, or the fulfilment of ease, articulates not general models of animated labour, but the subjective experience of work, its efforts and rewards. Hume's writing, in particular, paints a striking picture of the mental demands of his work, and its emotional and even somatic effects; Smith's concern with the philosopher is less personal, more theoretical, but nevertheless links philosophical labour with its emotional consequences through aesthetic and imaginative pleasure. In the context of philosophical work, then, Hume and Smith are concerned less with the mechanisms which prompt the subject to work, than with how work is itself experienced by the subject: as fulfilment, pleasure, effort, isolation. To consider the philosopher's work for each leads to and opens out the description of a subjective, inward space of consciousness, emotional experience and, variously, aesthetic pleasure or mental suffering. Rather than work being something to which one is subject, and which is caused by the mechanical forces of one's passions, or a constant, inescapable urge, work for the philosopher, according to Hume and Smith, offers an experience through which one might know one's subjective self. This is precisely the opposite of the self-alienating effects of the 'division of labour' which Smith documents elsewhere, where work operates to make one less, not more, human; such alientating effects are very different from Hume's experiences of the mental demands of work. For the philosopher, work is presented less as a consequence of how one is constituted as a subject, more as that through which one's own nature as a subject might be experienced and known.

Describing Pocock's account of arguments for and against the new commerce in the early eighteenth century, Furniss suggests that each side of the debate offered 'contending theories of history which implicitly adopted different sets of aesthetic assumptions and models of human subjectivity'.[34] To study the models of human nature offered by Locke, Hume and Smith is to see how a concern with the human personality, passions and psychology is not limited to the political debates over the value and significance of commerce, but informs, at a fundamental level, philosophical theorisations of economic activity. Accounts of what, after Smith, comes to be known as political economy, clearly involves the construction and modelling of human nature and the human subject in very particular ways: as subject to desire, as motivated by unease, as seeking ease and satisfaction, and as willing to labour to achieve his ends – characteristics shared by the three accounts considered above, whatever other variations exist between them. Writing on economic topics is thus not distinct from conceptions of what it is to be human, but arguably constructs and depends on a very particular understanding of the human, whether the terms of that understanding are derived, as with Hume, from the systems and models of a mechanical natural philosophy, or, as with Smith, from a more organically oriented language. The very language and sources of such models of human nature, however, might give us pause before we agree with Furniss that these writers were concerned with 'models of human subjectivity', a term which suggests inwardness, consciousness, privacy and individuality. Setting their accounts of their own labour as philosophers aside, Hume and Smith map such an inward space of subjective consciousness very minimally; although concerned with the passions and psychology of the economic subject, their interest is limited to discovering the 'secret spring', the animating mechanism which puts the human machine into motion, and the language of efficient causes which they adopt to that end itself mitigates against a descriptive detour of the consciousness or experiences of the subject. The introduction of his *Treatise* announces Hume to be interested not in an inner and individuated 'subjectivity', but in the more general concern of 'human nature', which is constructed as a knowable object, static, constant, observable: such a proper object of scientific study, in fact, as his own analogy with Newtonian natural philosophy would suggest, and one knowable via a division into its respective parts, of the understanding, the tastes and the sentiments, rather than through an exploration of inwardness, consciousness and individuality. Such an understanding of human nature, in Hume and elsewhere, as an object 'knowable' through properly analytical study, also establishes what is for him and others a very particular narrative of the human passions, psychology, desires and motivations as demonstrable, reliable and authoritative: a proper bedrock on which to secure a further layer of argument about political economy itself. Finally, the concern in these accounts with human nature – with the 'nature' of what it is to be human – leads on to the construction of political economy itself as a 'natural' entity, a process which, as the next section shows, is also achieved by borrowing and exploiting a language of the 'vitalist' body which the interest with the human subject in these texts has already introduced.

The physiology of political economy

Smith's economic writing is preoccupied with the human body not only as a system in whose 'springs' of action the hidden origins of labour, a constitutive clog in the watchwork of the economy, might be found, but also as the source for another dominant metaphor in his presentation of political economy, which is periodically figured in *Wealth of Nations* through bodily images.[35] The word 'economy' itself, or 'œconomy', as Smith spells it, might already carry suggestions of the bodily: it is defined by Johnson in 1755 as referring to the organisation, disposition or regulation of something, perhaps, as in Johnson's examples, a family or household, but the body was itself known as the 'animal œconomy', a term which also named the study of the organisation and functioning of living bodies until the use of the modern alternative, 'physiology', emerged at the end of the eighteenth century. The body had, of course, been a suggestive presence in political and economic writing since at least the seventeenth-century discourse of a body politic, which, inspired by Harvey's discovery of the circulation of the blood in 1628, reinvigorated a much older exploitation of the body as a rich source of metaphor for discussing the state or nation.[36] The rhetorical trick of figuring political subjects as incorporated and united into a single body transforms those who might be considered as radically at odds into constitutive parts of a unified whole, achieving an expression of corporality or aggregation in the face of seeming difference, and implying, against evidence to the contrary, the existence of shared, communal interests. Such corporeal unity is certainly at stake in one early eighteenth-century variant of this theme, Defoe's tale of the 'monster of Glasgow', where a deformed version of the body politic is used to illustrate the dangers of conflict between trade and land interests:

> Land and Trade are like the Monster of *Glasgow* ...; it had one Body from the Navel downwards, but two Bodies from the Navel upwards – They had different Hands to work, different Heads to Contrive, and, *no doubt,* different Souls to Direct; they receiv'd Nourishment two different Ways, and had two Stomachs to digest, but they had but one pair of Legs to walk with, one Belly to receive and vent, one Receptacle; and from hence it follow'd that they had but one and the same Life.
>
> The Foolish Creature would sometimes, just *like our Landed Men and Trading Men,* Quarrel with itself; one Side would be for going this Way, and the other that, an Evidence it had two Wills – And *What was the Consequence?* why the Legs were fain to stand still till the Heads were agreed, for there being but one pair of Feet, and the Locomotive Faculty receiving its orders from the Will – and there being two Wills, till they concurr'd, the Legs were perfectly Useless.
>
> Would to God our People would consider how apt this Creature was form'd to describe our Case; Really, good People, if Trade and Land, which are the Wealth of this Nation, are divided and differ, the whole Body will soon stand still – And this, like the Circulation in the Body, will throw the whole into Appoplexies, dead palsies, and every Mortal Disease.[37]

Writing at a time when the different interests of the old landed order and the new commercial enterprise were well established and vociferously rehearsed, Defoe's insistence on their compatibility is both optimistic and striking. Paradoxically, however, the figuration of those divergent interests as a monstrous deformity only makes the possibility of their equal incorporation into the national body seem even more remote: the body Defoe pictures here, with its doubled heads and organs, speaks more powerfully of a nation caught between two opposed 'Wills' than of a possible subjugation of opposing, lesser interests into the larger, shared interest of the nation. The transformation of difference into unity promised by the idea of the body politic could hardly seem more distant.

Defoe's image of the Glasgow monster works by positing a deformed and aberrant body where a more healthy, more perfect one should be. Its emphasis is on the 'Nation' as a 'Body', and, aside from a brief gesture towards the possibility of ill-health – 'Appoplexies, dead palsies, and every Mortal Disease' – the specifics of bodily operation and physiological detail are eschewed. Smith's concern with the body in *Wealth of Nations* is at once more precise and more extended, as he draws on the body both as a source of imagery for his arguments about the natural and 'healthy' operation of specific aspects of economic activity, and to depict the economy itself in a way which resonates with contemporary theories of physiology or the 'animal œconomy'. Initially, references to the body appear to continue a conventional imaging of economic activity through bodily physiology: Smith, like other writers before him, refers to the channels, organs and vessels of commerce, the circulation of money, or seeming 'symptoms of decay' in the trade of other countries.[38] It becomes apparent, however, both that Smith exploits the recurring analogy between the physiological and the economic to make specific points about wrong-headed economic management, and that his depiction of the body depends on notions of its own, innate, health-preserving principles – notions which add force to his frequent arguments about political maladministration. His discussion of the colony trade is one instance which illustrates how an economico-political argument against heavy interference in commerce and industry operates by exploiting an extended analogy with a body best left to its own health-promoting interests:

> The monopoly of the colony trade besides, by forcing towards it a much greater proportion of the capital of Great Britain than what would naturally have gone to it, seems to have broken altogether that natural balance which would otherwise have taken place among all the different branches of British industry. The industry of Great Britain, instead of being accommodated to a great number of small markets, has been principally suited to one great market. Her commerce, instead of running in a great number of small channels, has been taught to run principally in one great channel. But the whole system of her industry and commerce has thereby been rendered less secure; the whole state of her body politick less healthful, than it otherwise would have been. In her present condition, Great Britain resembles one of those unwholesome bodies in which some of the vital parts are overgrown, and which, upon that account, are liable to many dangerous disorders scarce incident to those in which all the parts are more properly proportioned. A small stop in that great blood-vessel, which has been artificially swelled beyond its natural dimensions, and through which an unnatural proportion of the industry and commerce of the country has been forced to circulate, is very likely to bring on the most dangerous disorders upon the whole body politick.

By contrast, a 'rupture' in trade with 'any of our neighbours upon the continent' would have rather less dramatic consequences:

> The blood, of which the circulation is stopt in some of the smaller vessels, easily disgorges itself into the greater, without occasioning any dangerous disorder; but, when it is stopt in any of the greater vessels, convulsions, apoplexy, or death, are the immediate and inevitable consequences ...

Some moderate and gradual relaxation of the laws which give to Great Britain the exclusive trade of the colonies ... can by degrees restore all the different branches of it to that natural, healthful, and proper proportion which perfect liberty necessarily establishes, and which perfect liberty can alone preserve.[39]

British trade with the colonies, Smith argues here, encouraged by various protectionist measures, has upset a more 'natural' balance which, left to its own devices, would characterise the body politic. Like Defoe's 'Glasgow monster', Smith's emphasis is on a deformed and grotesque body; but whereas with Defoe it is the perhaps inevitable competition between different interests which have disrupted the healthy unity of the nation, here it is government mismanagement, in the shape of ill-conceived economic policy, which has caused the threatened 'disorders': the impulse of the body itself is to 'that natural, healthful, and proper proportion' which in a state of 'perfect liberty' would 'necessarily' be established. If Defoe's monster works as a parable to quieten, or at least harmonise, the competition between land and trade, Smith's imagery here sees the pursuit of different interests, in the 'different branches' of trades, as a healthy function of the body politic which the administration should encourage.

Smith's discussion of the colonial trade operates by opposing the pursuit of trade, industry and commerce, characterised as essentially vital, natural and healthy, with political misadministration. This perception of national economies as healthy despite government interference recurs throughout the text. Discussing taxes, 'a curse equal to the barrenness of the earth and the inclemency of the heavens', he notes that, just as 'the strongest bodies only can live and enjoy health, under an unwholesome regimen', only the strongest nations can prosper under heavy tax regimes, and prosper not 'by means of them' but 'in spite of them'.[40] The establishment of any kind of monopoly 'introduces some degree of real disorder into the constitution of the state, which it will be difficult afterwards to cure without occasioning another disorder'.[41] Mercantilist doctrines, Smith's targets in all these examples, meanwhile, are described as the tenets of 'pretended doctors', who attempt to cure, but only further harm their patient by their interference.[42] Given such widespread evidence of bad economic medicine, Smith's readers might be forgiven for asking after the cause of the increasing prosperity of their own country, at least, and Smith's suggestion that their own efforts in pursuit of their self-interests secure national prosperity is also figured in bodily terms:

The uniform, constant, and uninterrupted effort of every man to better his condition, the principle from which publick and national, as well as private opulence is originally derived, is frequently powerful enough to maintain the natural progress of things toward improvement, in spite both of the extravagance of government, and of the greatest errors of administration. Like the unknown principle of animal life, it frequently restores health and vigour to the constitution, in spite, not only of the disease, but of the absurd descriptions of the doctor.[43]

The constant effort in pursuit of self-interest, which we saw earlier animates the labour of the worker, is revealed here as the hidden 'spring' also of the health and vitality of the body politic. That secret effort, like the secret origins of life itself, the 'unknown principle of animal life', animates also the life of the political body, preserving its health in an unintended way, and despite the 'doctor's' harmful prescriptions.

This imaging of the worker as the hidden life-force of the economy integrates two identifiable levels of Smith's discussion in *Wealth of Nations*: his 'microeconomic' concern to understand the 'springs' in human nature which prompt man to labour, and his 'macroeconomic' interest in mapping the interrelations between different forms of economic activity and their contribution to the entity of the economy itself. 'Micro-' and 'macroeconomic' here might also be expressed in terms which Smith himself would have recognised: his concern with the passions and psychology of man corresponds to the Scottish Enlightenment's 'science of man', whilst he himself labels his mapping of the correspondences and interconnections between different aspects of man's economic activity as a new study of 'political œconomy'. Identifying these two levels demonstrates the sustained and complex nature of Smith's 'body politic' image, which operates at both levels, figuring labour as the 'vital principle' of the economy whilst fleshing out and animating the economy itself as a health-preserving body, and in addition supporting Smith's economically grounded arguments against mercantilist policies. It is also an image which appears to deploy a quite specific understanding of the nature of the body. Defoe's 'Glasgow monster' is little more than a momentary, sketched grotesque, but Smith's bodily analogy repeats, highlights and exploits a particular perception of contemporary medical science: the idea of unknown, but nevertheless powerful, innate health-preserving principles in the body, which unconsciously pursue the body's 'best interests' by maintaining and regulating its healthy state, and by attempting where possible to return it to health when this was endangered. This perception resonates in Smith's reference, in the passage quoted above, to the 'unknown principle of animal life' which 'restores health and vigour to the constitution, in spite, not only of the disease, but of the absurd descriptions of the doctor', and it is repeated in an attack on the French Physiocrat François Quesnay and his contention that the political body would prosper only under a 'certain very precise regimen … of perfect liberty and perfect justice'. Quesnay's mistake, Smith says, is analogous to that of those speculative physicians who claim that 'the health of the human body could be preserved only by a certain precise regimen of diet and exercise'; in fact, 'experience would seem to show that the human body frequently preserves, to all appearances at least, the most perfect state of health under a vast variety of different regimens; even under some which are generally believed to be very far from being perfectly wholesome'. It would seem that the 'healthful state of the human body ... contains in itself some unknown principle of preservation, capable either of preventing or of correcting, in many respects, the bad effects even of a very faulty regimen', and this same principle of preservation in the political body, Smith goes on, is 'the natural effort which every man is continually making to better his own condition ... a principle of preservation capable of preventing and correcting, in many respects, the bad effects of a political œconomy, in some degree, both partial and oppressive'. Just as the 'wisdom of nature' has made 'provision' for the health of the 'natural body', so too is the health of the 'political body' secured by the same 'principle of preservation': man's 'natural' pursuit of his interests.[44]

Smith's conception of the health-preserving principles of the body in fact reflects both current ideas in medical practice and new developments in physiological theory taking place in Edinburgh at mid-century. These might best be approached by considering advice given by Smith to the severely ill Hume in 1776 – the very year *Wealth of Nations* was published. In June of that year, struggling with the illness which was eventually to kill him, Hume travelled to Bath to take the waters, but experienced no significant improvement in his condition. In correspondence, Smith suggests that the waters may have interfered with the natural efforts which his body would already have been making to restore itself to health:

> A mineral water is as much a drug as any that comes out of the Apothecaries shop. It produces the same violent effects upon the Body. It occasions a real disease, tho' a transitory one, over and above that which nature occasions. If the new disease is not so hostile to the old one as to contribute to expel it, it necessarily weakens the Power which nature might otherwise have to expel it.

According to Smith, any improvement in his health which Hume had noticed had an altogether different source:

> Change of air and moderate exercise occasion no new disease: they only moderate the hurtful effects of any lingering disease which may be lurking in the constitution; and thereby preserve the body in as good order as it is capable of being during the continuance of that morbid state. They do not weaken, but invigorate, the power of Nature to expel the disease. I reckon it probable that the Bath Waters had never agreed with you, but that the good effects of your journey not being spent when you began to use them, you continued for some time to recover, not by means of them, but in spite of them.[45]

In recommending the effects of 'change of air and moderate exercise' on the 'power of Nature' to expel disease, Smith may be remembering advice given by his own doctor, William Cullen, who in July 1760 told Smith 'that if I had any hope of surviving next winter I must ride at least five hundred miles before the beginning of September'.[46] Cullen's remarkably moderate and limited intervention in the face of a frankly disheartening prognosis again suggests a reliance on the self-healing capacity of the body, whose restorative forces would be stimulated by the same air and exercise which Smith later recommends to Hume. As in his denouncement of mercantilist policies in the *Wealth of Nations*, Smith's advice to Hume, like that he received from his own doctor, rests on faith in an innate, curative 'power of Nature' to promote the body's self-health and self-preservation, against the potentially damaging prescriptions of less enlightened doctors.

In the context of contemporary Scottish medical practice, Smith's doctor William Cullen was a prominent figure, an influential teacher and practitioner whose work reflected some of the new physiological theories being produced by the Edinburgh Medical School to which, from the mid-1750s, he belonged.[47] Particularly through the work of Cullen's colleague Robert Whytt, the Edinburgh Medical School was at the forefront of attempts to develop new, 'vitalist' physiological theories, which sought to replace the largely mechanistic models of bodily action which had been dominant since the late seventeenth century with new theories capable of understanding the body not in terms of laws of motion or other mechanical models but as a specific life-system. Such theories depended on an alternative conception of the body as a 'natural' entity, having its own internal forces and animating energies, through which it could be understood to be independently self-regulating and self-preserving. Such 'vital forces', which were thought responsible for such actions as involuntary motion, the operation of the vital organs and other essential but unconscious or automatic actions, were envisaged differently by various theorists: William Porterfield at Edinburgh offered an animist account, whilst Théophile Bordeu at Montpellier formulated a theory of an unconscious 'sensibility' transported by the nerves, which later informed the Encyclopaedists' descripton of sensibility.[48] Such theories went hand in hand with new research into the nervous system, thought to be the location of the body's animating principles. Research into the perceptive and responsive capacities of the body, the ability of muscles to respond to stimuli, or the role of the nervous system in co-ordinating muscular movement, produced a conception of the body as a unified whole, capable of organised, automatic, autonomous responses in pursuit of its best interests, a conception which challenged an established, Cartesian conception of a body as a machine whose movements were governed by the conscious, rational mind.[49]

Such vitalist physiology was, from the 1730s, well established at the Edinburgh Medical School, and contributed to the development of a characteristically Scottish practice of medicine, especially in contrast with an

English physiology, which continued to make reference to a mechanical model of the body.[50] Edinburgh professor William Porterfield's publication, in 1737, of a vitalist account of eye movement, could be seen to inaugurate two decades of research by himself and his colleagues which contributed to the development of a vitalist account of the body and the nervous system. The most significant figure at Edinburgh, however, was Robert Whytt, who suggested the existence of a 'sentient principle' located throughout the body in the nervous system, which caused bodily sensitivity, controlled muscular action and determined the responses of the body to external stimuli.[51] His work, including researches on the relationship of the muscles to the nervous system, and a theoretical justification for the 'vital' or sentient principle, was presented to the Edinburgh Philosophical Society, to which both Hume and Smith belonged, and he also gained prominence through a long-running dispute with the influential Albrecht von Haller over the exact extent of the sensibility of the body and the role of the nervous system in controlling muscular action.[52] Through the sentient principle, Whytt gave an integrated picture of the body's actions, understanding its different parts to communicate via the nervous system, and seeing the whole as acting necessarily, involuntarily, unconsciously and spontaneously in response to stimuli. Such a mode of action was especially marked in the body's response to any threat, challenge or pain, when it could be observed to pursue, unconsciously and involuntarily, the 'wisest' course of action to promote its own self-preservation and ease. Although Whytt at times referred to the sentient principle as the soul, it is clear that his theory understood the body to be an autonomous and self-regulating system, capable of pursuing self-government and self-preservation, independently of any control from an overseeing, rational mind.

The 'vital principle' was Whytt's contribution to what was at stake in the new vitalist physiology: the development of an understanding of the means of connection and communication between the different parts and organs of the body, other than according to mechanical terms, or by reference to some separate, overseeing, governing agency. As with other 'vitalist' versions of animal œconomy, his theory figures such bodily communication and direction in terms of the energies and forces already inherent within the body, and constructs an account of the body as a living system fitted to unconsciously, but ably, pursue its best interests by determining its own responses to external stimuli. The 'vital' principle plays a crucial part in this, co-ordinating communication between different parts of the body in order to promote the body's best interests through, for instance, reflex or involuntary action to diminish pain or increase ease. The supposition of the vital principle thus makes it possible to conceive of the body as a self-directing organism, following the 'wisest' course of action involuntarily, unconsciously and independently of any external intervention, such as that of a rational mind or will. The expression by Smith – who, in an early essay on the natural philosophy of the ancient Greeks, refers twice and unquestioningly to the 'Vital Principle' which animates the life of both 'plants and animals' – of the individual's self-interest as an 'unknown principle of preservation', highlights how, in a similar fashion, actions undertaken by an individual in pursuit of his interests also constitute a means by which the 'body' of the economy is controlled, regulated and preserved.[53] Explicitly, we are told that such pursuit of his interests preserves national prosperity even in the face of government extravagance or mismanagement, but other instances recur frequently throughout the text in which the actions of individuals, acting for themselves, unconsciously promote or preserve the health of the body politic. These might include the transference of labour and investment from less profitable to more profitable trades, the effects of competition to lower or raise prices, or, in the labour market, the withdrawal of underpaid labour in favour of other, better-remunerated work elsewhere. Such actions, of adjustment and response to economic reality, at the micro level, operate like the self-regulatory efforts of the vitalists' animal œconomy, through which it preserves its health and maximises its wellbeing. In their effect in promoting the health of the economic whole, such actions are as unconscious, yet co-ordinated, as the involuntary responses of the nervous system; like them, their combined effect is beneficial to the larger system; and like them again, they are not controlled by an external, overseeing, governing function. As with the development in physiology of alternatives to mechanical models of the body, Smith replaces a traditional model of an externally governed economy with an integrated, autonomous, self-regulating system, whose unconscious pursuit of its best interests should be allowed to operate without interference.

Smith's language of 'natural effort' and 'principles of preservation', then, highlights the extent to which his economy is itself a 'living system', animated by the labours of the individuals who promote its healthy prosperity. His physiological imagery expresses an understanding that the wealth of the nation is the unintended consequence of the cumulated activity of labouring bodies: the economy, according to Smith, is not a cause of oppression to human effort and action, but a beneficial, unconscious, effect created by the efforts which men are always already making. Such a perception, of course, was lost on later, nineteenth-century critics of political economy, who characterised it as mechanical, non-human and machine-like. Smith's conception of the connection between economic subjects and the economy – the latter as created and animated by the desires and efforts of the former – also offers a new, optimistic formulation of the relation between subject and nation. Mandeville's troubling equation of private vice and public benefit had suggested that national prosperity was founded on the ultimately immoral passions of avarice, selfishness and desire, an equation which, whilst linking, as Smith does, the passions and actions of the individual to the national whole, can only understand the prosperity of the latter in terms of the sufferings, or moral weaknesses, of the former. This perception is present too in Mandeville's depiction, like Smith's, of the economy as a living being, but one whose life-force is not provided by the efforts of individual self-betterment, but by the inevitable vicissitudes of fortune:

> Philosophers, that dare extend their Thoughts beyond the narrow compass of what is immediately before them, look on the alternate Changes in the Civil Society no otherwise than they do on the risings and fallings of the Lungs; the latter of which are as much a Part of Respiration in the more perfect Animals as the first; so that the fickle Breath of never-stable Fortune is to the Body Politick, the same as floating Air to a living Creature.[54]

Mandeville's image here figures the ups and downs of individual fortune as a natural process, inevitable and necessary to the life of the body politic, viewed merely with a philosophical acceptance by those capable of a larger view. In his image, the perception of the body politic as a 'living Creature' simply implies one's necessary, passive submission to the larger entity of which one is part, the life of which one's own fortunes, whether good or bad, will momentarily sustain. In his *Theory of Moral Sentiments*, Smith is strongly critical of the stoicism which is implied here, according to which one must accept one's place as 'an atom, a particle, of an immense and infinite system, which must and ought to be disposed of, according to the conveniency of the whole'.[55] His own formulation of the body politic as animated not by mere changing fortunes, but by the active pursuit of individual prosperity, reshuffles the image Mandeville presents. Subjects are not passive participants in a larger body to which they inevitably belong, but active promoters of their best interests; such individual efforts cumulatively, though unconsciously, create the economic body, through which their efforts are connected, and which is not immune to their fortunes but itself sustained by them. Smith's body politic, unlike Mandeville's, is animated by the 'vital' principle of self-preservation; his subjects exert themselves for benefits which are at once particular and general, private and public. If, for Mandeville, economic subjects inevitably belong to a body sustained by but impervious to their prosperity or otherwise, Smith's alternative physiology of the body politic conceives of the relationship between subjects and nation as harmonious, integrated and mutually beneficial.

The language of the body deployed in *Wealth of Nations*, then, is not mere rhetoric, but invokes a specific context of medical and physiological knowledge in Scotland at mid-century. That knowledge is applied to an equally specific need to satisfactorily formulate the relationship between economic subjects and the body politic in the emergent commercial nation of eighteenth-century Britain, where growing recognition of the value and power of trade and commerce was accompanied by anxieties about its moral cost, social effects and political consequences. The language of vitalist physiology contributes in important ways to Smith's theorisation of the economy, through imaging the specificity of the kind of body the economy might be, including suggesting its modes of automatic functioning, and by supporting Smith's arguments against both mercantilist doctrines and other forms of interventionalist government policy. The physiology of Smith's economy thus also implies a politics: it can be deployed effectively to counter certain notions of government economic management and administration by offering an alternative account of the national 'body', whose life and well-being are generated by its constitutive subjects, rather than its political masters. Within the confines of *Wealth of Nations*, and its potentially delimited concern with economic management, the larger political implications of conceiving human activity in terms of an independent, self-directing, nature which is best left to its own devices, are not too evident, and are in any case readily contained. As the next chapter demonstrates, however, in the context of more wide-ranging, explicitly political debates on human society and governance such as those to which John Thelwall contributed in the 1790s, such a parallel with a materialist, self-governing, self-determining nature, is both more explicitly, deliberately drawn, and rather more politically explosive.

Part II
Enlightenment in the 1790s: The Scottish Legacy

4
Enlightenment Legacies and Cultural Radicalism: Physiology and Politics in the 1790s

> one would have thought that the existence of theological and political institutions had depended upon the agitation of a question in physics
>
> John Thelwall, Prefatory Memoir, *Poems, Chiefly Written in Retirement* (1801), p. xxiii

Introduction: vitality and reanimation, 1777–89

On 27 June 1777, Dr William Dodd – society preacher, 'Macaroni' and one-time chaplain of the Magdalen Hospital for reformed prostitutes – was executed for a forgery committed in a fraudulent attempt to borrow money on behalf of his ex-pupil Philip Stanhope, Earl of Chesterfield. Until his final moments, Dodd had hoped for a pardon from the king to whom an appeal, written in his name by Samuel Johnson, had been sent, but despite numerous petitions and newspaper campaigns, no last-minute leniency materialised. Indeed, the vociferous appeals on behalf of Dodd may have had the opposite effect, in deciding Lord Chief Justice Lord Mansfield against setting a dangerous precedent in allowing public opinion to influence the operation of justice. A speech (also penned by Johnson) to be declaimed by Dodd from the gallows went undelivered, and the clergyman was hanged for about an hour before his body was taken down and carried away.

Whilst this was the end of one story, it was just the beginning of another. For the next half-century, claims and speculations about Dodd's possible afterlife, or return to life, were to be traced intermittently through the pages of the fashionable periodicals of London, Scotland and the north of England, to intrigue and fascinate the reading public. In one version of this sequel, Dodd was taken from the Tyburn gallows to the house of Mr Davies, undertaker, where a waiting group of medical men, including a Mr H and a Dr C, after several effortful hours, succeeded in reviving him. In September 1777, the *London Review of English and Foreign Literature* reported an Irishman's claim to have dined with Dodd at Dunkirk after his supposed death, a claim he was ready to support by oath.[1] In 1794, some 17 years after Dodd's putative death, the *Aberdeen Journal* published an account, derived from the papers of a recently deceased 'Gentleman of Glasgow', of Dodd's resuscitation. Mr H and a Mr D, it claimed, had 'stripped and exercised friction' on Dodd's body for two hours after his death, when at last 'they perceived a motion of his breath', followed by a 'sweat that spread itself over the Doctor's body, and a continual panting and groaning'. The pain which accompanied Dodd's return to life was so great, it commented, that he felt his life was 'hardly worth purchasing at so dear a rate' – but purchase it he did, as this account at least would have it.[2] The credibility of such claims, which appeared to stem from authoritative sources, was all the greater given the absence of any reports or public knowledge regarding the circumstances of Dodd's burial.

Claims and counter-claims about Dodd's death, or life, were evidently as long-lived as some of the stories would have Dodd himself be, for the matter was still generating interest nearly 30 years later. In 1822, 'Onesimus' wrote to the *Newcastle Magazine* referring to the 'great many gossips' stories' still circulating about Dodd and appealing for any new information. His letter elicited a reply from Charles Hutton, mathematician and Fellow of the Royal Society, whose 'positive statement of the truth' apparently puts an end to the trail of speculation. One of a number of Royal Society Fellows who used to retire to Slaughter's Coffee House in St Martin's Lane to eat oysters and 'hold familiar discourse' after the more formal weekly Royal Society meetings, Hutton recounts how on one occasion discussions turned to the notorious Dodd case – a topic seemingly as fascinating for scientists as for gossipy periodicals. John Hunter – the 'Mr H' of more coy accounts – who was present and 'known to be one of the medical gentlemen who attempted that experiment and who was, indeed, the chief operator in it' then offered a full account of the episode, which Hutton now reported. Stating that the 'brethren' had come together with the purpose of attempting to recover Dodd's 'animation', Hunter revealed that they 'tried all the means in their power for the reanimation' but without success, and that they considered the chief cause of their failure the length of time taken to transport Dodd's body from Tyburn, which had substantially delayed the start of their 'experiment'.[3] Outlandish as the reports of Dodd's resuscitation which had circulated for the previous 50 years might have seemed, they had their roots in an 'experiment' performed by one of the leading physiologists and anatomists of the last decades of the eighteenth century. The perennial cultural fascination with the possibility of reanimation shown by the long-lived appetite for news of Dodd's revival echoed, in a different vein, a similar curiosity about resuscitation on the part of a leading experimenter of the day.

Perhaps unsurprisingly given its highly controversial nature, Hunter never commented more publicly on his 'reanimation' experiment, but focused as it is on central questions of the nature of life, death and 'animation', his involvement with Dodd's case correlates with his more readily acknowledged preoccupations pursued over a lengthy career as a physician and experimenter. A Glasgow-born autodidact who followed his older brother, William Hunter, into a successful London career as anatomist, lecturer and medical man, Hunter was a man of seemingly eclectic interests but many talents. Employed first by his brother as a lowly dissector and lecture demonstrator, his skills soon led him into independent researches into the anatomy and physiology of human and animal bodies. Hunter's interest in comparative anatomy led him to amass a vast and expensive collection of exotic animals who shared his increasingly cramped family home just outside London at Earl's Court, and numerous animals, exotic and otherwise, were submitted to endless experiments as Hunter tirelessly pursued his quest for a fuller understanding of human and animal life, the nature and operation of bodies, and the causes of disease and death. In one of many such experiments, Hunter froze fish and rabbits' ears, hoping to reanimate them when thawed, and to make his fortune by using the same method to extend mankind's life to lengths previously unimaginable.[4]

Hunter's fish-freezing experiments were only one way in which he pursued his interest in what he termed the vital or 'life principle', a concern which he claims in his *Lectures on the Principles of Surgery* to have occupied him for over 30 years and studied through numerous experiments. This principle emerges in Hunter's attempt – which had occupied Robert Whytt and other vitalist physiologists before him – to differentiate between animated and unanimated, or 'common' matter. His brother, William, in his own anatomy lectures – one course of which was attended by Gibbon, Edmund Burke and Adam Smith – had already noted the 'superiority' in 'the natural machine, the animal body' over 'machines of human contrivance or art', in the 'internal powers of the machine itself', by which healing, self-preservation and other actions take place. The elder Hunter fulminated against those who, instead of 'making accurate observations upon animals themselves', wasted time on 'mechanical and chemical visions' – seeking to explain such animal function by reference to the laws of mechanics or chemistry, rather than recognising the specificity of whatever mysterious powers of life operated in animated bodies.[5] Such convictions, which echo those expressed by Whytt and others, were perhaps acquired during his medical education at Edinburgh. Whilst William Hunter recognised the existence of a 'principle' regulating life, his brother went further in attempting to identify or theorise the unknown 'something' which put matter into motion, and to elaborate its various powers. He had first noticed the superiority of the healing powers of nature over surgical interventions whilst serving as an army surgeon in Portugal in 1762, and his most extended account of the body's internal life-force is given in his *Treatise on the Blood, Inflammation, and Gunshot Wounds* (1794) which draws on that experience.

A central claim of Hunter's *Treatise on the Blood* was that 'mere organization can do nothing, even in mechanics'; rather, matter 'must still have something corresponding to a living principle; namely some power'.[6] Whilst he considered that this life-force was located throughout the body, in even the smallest particle, as a cause of every action and a property of every part, he gave blood a leading role in transmitting and maintaining the living power of the body. It was blood, Hunter asserted, that gave the body, in all its parts, the power of preservation, the susceptibility to impressions, and its ability to act and to react – all actions which collectively enabled it to resist disease, restore itself to health after injury and undertake other vital acts. The same life powers were observable within blood itself, which, Hunter noted, could coagulate when required – a 'spontaneous change' which was 'necessary to the growth, continuance, and preservation of the animal'.[7] For Hunter, the living principle, in short, was an internal power causing and manifested in the numerous physiological functions performed by organic structures in their self-maintenance and self-preservation. Central to life, because no physiological action could be understood without reference to the living principle as cause, it was both a necessary theoretical assumption for physiologists keen to distinguish the properties of organic from inorganic matter, yet also mysteriously aloof – everywhere observable, nowhere fully knowable. Before the discovery of oxygen, as well as of other more particular physiological functions, reference to the 'living principle' enabled Hunter and others to construct a coherent and complete account of human and animal bodies as living entities – although at the cost of some controversy over the precise nature and location of 'life'. Hunter's own recognition of the ways in which his identification of a crucial life-force with the blood echoed ancient Judaic teachings recorded in the Bible did not prevent others from seeing dangerously materialist tendencies in a doctrine which attributed the highest powers of life not to God but to what appeared to them to be organised matter.

Hunter's belief in the animating powers of the body, whilst at one level a theoretical assumption, inevitably informed his practice as a physician. This is already shown by his involvement in the extraordinary attempt to revive Dr Dodd – an attempt which depended on the attempt to stimulate the body's latent life-forces to a renewed activity. The methods taken by Hunter in his 'experiment' on Dodd in 1777 were probably similar to those recommended by him the previous year in his 'Proposals for the Recovery of People Apparently Drowned', a paper presented to the Royal Society in March of that year, and later published. Announcing that the subject was 'closely connected' to the enquiries on the 'loss and recovery of life' which 'for many years, have been my favourite business and amusement', Hunter asserted that drowning represented not a destruction but only a 'suspension' of the 'actions of life', and that recovery of such persons should aim to revive their 'powers of life' by the applications of medical art. Referring to the 'living principle' inherent in the blood, 'which preserves the body from dissolution with or without action, and is the cause of all its actions',

Hunter outlined various treatments which, in diverse ways, aimed to 'rouse the living principle'. These include the use of stimulating medicines, the blowing of air into the lungs, warming through bed clothes and steam (as heat is 'congenial with the living principle'), and the avoidance of cold and bleeding due to their weakening effects.[8] A year earlier, as discussed in the Introduction, William Cullen, teacher and correspondent of John Hunter's brother, had made similar recommendations in his own publication on this topic, aiming, like Hunter, to restore the 'vital principle', defined for him as a 'certain condition of the nerves, and muscular fibres' on which the 'living state of animals ... especially depends'.[9] Clearly one medical effect of belief in the living or 'vital' principle was a perception that the boundary between life and death might be crossed in both directions.

If John Hunter's career was devoted to the investigation of the nature of life, its loss and recovery, many others were prepared to share his belief that knowledge of a 'living principle' made recovery from death a possibility – as well as to exploit the sentimental effects and moral suggestions of such a recovery. As well as being a topic for serious scientific research and a subject for speculative gossip, reanimation had a significant cultural presence in the late 1770s and 1780s. Hunter's 'Proposals for the Recovery' were written in response to a request from William Hawes, who three years earlier had founded the Humane Society, devoted to the rescue and recovery of drowned persons; similar organisations had recently been founded in Continental Europe. Ironically, Dr Dodd had also been involved in establishing this charity, for whom he preached an anniversary sermon in 1776. The Humane Society reports offer page after page describing precisely such resuscitations as Hunter had attempted on Dodd, and its *Transactions* identify Cullen and Hunter as leading authorities on the topic, reiterating their recommended actions to restore the 'vital principle'.[10] Hawes, the same publication asserts, had been lecturing on 'suspended animation' since 1776, and interest in the 'principle of vitality' had even crossed the Atlantic to Massachusetts, where a Dr Waterhouse had given a discourse on that topic, defining vitality as heat or an electric fluid, and referring back to the Edinburgh vitalist Robert Whytt. Such fashionable charitable efforts no doubt contributed to the establishment of the possibility of recovery from the dead, in cases of drowning at least, in the public mind. When, in Maria Edgeworth's *Belinda* (1801), the benevolent Dr X successfully gives directions to 'restore Mr. Hervey's suspended animation' after he had ill-advisedly attempted to swim in the Serpentine after a morning spent drinking wine with some foppish friends, the event is not sufficiently remarkable to justify further narratorial comment, and speaks more of the potential of Hervey's moral recovery than any physiological exceptionality. Reanimation of the dead is also attempted in Mary Shelley's *Frankenstein* (including the attempt to revive the murdered Henry Clerval) – a text more usually discussed (as Chapter 6 will show) in the context of theories of the spontaneous generation of life.[11]

Vitality, reanimation and sentiment

A more extended association than Edgeworth offers of resuscitation with moral recovery occurs in John Thelwall's *Poems on Various Subjects*, published in 1787. Now known primarily as a political campaigner and lecturer in the fervour of the 1790s, Thelwall had an earlier career as a poet and writer. His *Poems* offers a range of verses on sentimental and pastoral subjects, with young love negotiating the obstacle of parental tyranny being a recurring theme. This topic receives its most developed treatment in 'The Seducer', the five-canto poem on which, the poet tells us, he has spent most of his efforts. A prefatory essay announces, conventionally enough, that the poem has been written in 'zeal for the united causes of virtue and beauty' and to display the 'modern vice of seduction' in its 'true colours', but it is with a particular aspect of the story of seduction that Thelwall is concerned. Parental anger, he asserts, adds greatly to the distress of a seduced child, and, if accompanied by an abandonment of their daughter (for it is always daughters, not sons, who are the focus for such moral anxieties), could lead to her fall into prostitution. As the sermons preached by Dr Dodd at the Magdalen Hospital for 'penitent' prostitutes suggest, depicting the distressing fate of seduced young women was a favourite means of stimulating the exercise of fashionable sensibility, and Thelwall's poem, even with its novel counsel of parental forgiveness, appears to offer the same desirable pleasures.[12]

But it is the means chosen by Thelwall to reinforce his emphasis on parental mercy which is especially noteworthy in the context of a broader cultural fascination with resuscitation and reanimation. By Canto 4, Amanda, seduced daughter of Sir Thudor, fleeing parental ire and personal misery, and pursued by her lover Damon and his friend Pastorus, runs into a wood, where she throws herself off a cliff, closely followed by Damon. Horrified, Pastorus hurries down to the cliff's foot in search of their bodies, only to discover the lovers alive, as though strangely resuscitated from a seemingly inevitable death – a theme emphasised, in the final lines of the canto, by an apostrophe to the 'parents, friends and lovers' of those reanimated by such means as practised by Hunter and the Humane Society:

> Ye, when the arts humane of pious men
> (Oh blest Philanthropy! thy agents here)
> Have wak'd the dormant spark of life again,
> And chang'd to transport horror's starting tear; –
> Ye, ye can guess, from what yourselves have felt,
> The mingled passions in his soul which dwelt.[13]

The amazement and confusion experienced by Pastorus, and Thelwall's readers, can only be understood by invoking the experience of those who have witnessed resuscitation's transmutation of horror to joy. By ending the canto with the apparent deaths of the lovers, Thelwall has already delivered a moment of great emotional crisis, but their supposed revival enables him to turn the screw of his readers' emotional response, transforming distress to relief and pleasure, and thus, by delivering such contradictory passions near-simultaneously, heightening and making more complex the sentimental experience offered by his poem.[14]

But Thelwall has further reasons for exploiting the reanimation topos. As Pastorus discovers in Canto 5, Amanda has in fact been saved by her father who, now living as a hermit in the wood, recognised her cry as she fell from the cliff and rushed from his cave in time to catch her. Damon, witnessing this, aborted his own cliff-top leap in its final moments. Amanda's 'return to life' from the shamed position of seduced woman is thus due not to the skilled medical treatment such as might be offered under the Humane Society auspices, but to the intervention of her previously hostile father: the power of renewed life stems not from physiological science but the impulses of paternal love. Resuscitation in Thelwall's poem is thus less a medical fact than a figuration of the effects of the same parental forgiveness exhorted in the prefatory essay. This representation of parental care as reanimation is made even more clear when Amanda's recovery follows attentions from her father and lover – kissing her lips, wiping her face, placing her hand in their breasts – which re-enact in romance mode the detailed physical treatments for resuscitation which were recommended by Hunter and Cullen. Such efforts, however, cannot prevent the eventual death which cultural logic so often assigned the seduced woman, and Amanda's death is soon followed by those of her grief-stricken father, who throws himself into the river, and of her lover, who dies a kind of spontaneous death as his body, enacting retributive justice on the moral transgressions of the flesh, delivers the proper 'vengeance' which 'on such crimes is hurl'd'.[15] Only Pastorus is left to make the sorry tale public and deduce the necessary morals.

So the deaths which reanimation – medical or moral – might have prevented take place anyway, although delayed enough to show both the redemptive powers of forgiveness (powerful enough to ensure Amanda's temporary recovery from her cliff-top fall) and the sinful nature of seduction, whose punishments nevertheless cannot, after all, be avoided. Resuscitation, medical or otherwise, cannot divert the 'stroke of fate' which the body's own moral economy enacts on the lovers.[16] But the play on the borders between life and death which is made possible through the poem's resuscitative themes articulates the moral and emotional landscape both of the poem and of the sentimental discourse on seduction to which it belongs, enabling father, daughter and lover to reach a moral understanding both of their own transgressions and of the punishments which, it is implied, justly meet a daughter's shame, a father's anger or a lover's seduction. Delayed, and reached not through wilful self-destruction but a timely ripening of fate, Amanda's initial 'resuscitation' ensures that the runes of her eventual death's full meanings can properly be read.

Resuscitation is even more prominent in the second extended poem in Thelwall's volume, 'A Dramatic Poem, founded on Facts, Recorded in the Reports of the Humane Society'. As the full title makes clear, the poem is closely inspired by the activities of the Humane Society, to whose patron, president and officers it is dedicated. Deploying the elevated literary form of blank verse, it offers extended sentimental narratives of lives curtailed and reanimated, based on facts reported in the briefest fashion to the Society by those applying for its rewards; so closely does Thelwall follow the Society's reports that he even gives relevant case numbers, as though anticipating a reader with both texts to hand. The relation of his poem to the charity both motivates and troubles the poet, however, who worries in one anxious footnote that, whilst he wished to publicise the variety of the work carried out by the Society, he has been unable to make the 'facts appear tolerably in a poetical dress'.[17] As his own method implies, his response to this perceived conflict between scientific fact and poetry is to downplay the former whilst maximising the moral resonances and sentimental effects which they might generate. As with 'The Seducer', resuscitation's interest lies less in its medical possibility than in the lessons of hope, redemption and recovery it might suggest.

This strategy of generating maximum sentimental narrative from the minimal case details offered in the Humane Society reports is most evident in the poem's central tale. Prompted by the Society's case no. 481, which in the barest details reports the recovery of a young woman who had hanged herself following 'cruel useage' by her seducer, Thelwall embroiders the extended story of Sophia, seduced and rejected by Roldan, and rebuked by her father, who hangs herself through guilt and shame. Pharmacinus, a pupil of the 'sage Heranicus' who is identified in a footnote as 'the lecturer on suspended animation', counsels against despair and tells of the 'wond'rous triumphs of Resuscitation' and its 'life-restoring art'. In support of his claims, other successful cases are narrated – again deriving from the Society reports – including a boy recovered from severe hypothermia, whom Thelwall presents as a 'friendless' waif in direct contradiction to the report's detailing of the attentions given him by his concerned companions. Emphasising the pitiful state of those on the receiving end of the resuscitators' attentions ensures that reanimation can be understood as another version of a well-rehearsed sentimental narrative, in which charity and benevolence are passively received by a deserving, distressed victim. In such a narrative, the 'art' of resuscitation can thus be another manifestation of social care given by their social superiors to the lower orders, and its potentially disturbing medical or theological ramifications contained and defused. Thus, when Sophia's brother Edmund is told of

the godlike art
Of rousing into life the dormant sparks

> Of animation, and the latent fire
> Rekindling with resuscitating breath
> Of Medical Benevolence,

the transition from 'godlike' to 'benevolence' transmutes an admittedly divine and therefore potentially disturbing power into an act of humanity.[18] Equally, of the details of physical states and medical treatments given in the Humane Society reports, it is only those which have a potential literary or sentimental resonance that Thelwall repeats in his poem. Sophia's own resuscitation is effected, disappointingly, off-stage, and the hypothermic boy is described only as receiving 'kind assistance', suggesting it is the larger figurative meanings, rather than the messily particular details of treatment, which most engage the poet. Whilst Amanda's 'resuscitation' had written large the importance of parental forgiveness, Sophia's recovery, swiftly followed by a reconciliation with her lover, suggests the possibility of an alternative fate for the seduced woman, in which socially transgressive sexual desires are legitimated by marriage.

In 1787, resuscitation narratives enable Thelwall to extend and rework received sentimental stories and tropes – the reformed seducer, the deathbed reconciliation – whilst enlarging on new moral emphases, such as parental forgiveness. Within a few years, as the next section will show, writing in an explicitly political register and advocating not familial harmony but political justice, Thelwall will exploit the figurative possibilities of reanimation in a very different way. That he is able to reuse a language and metaphorics of vitality and animation in such a different context points not only to Thelwall's own versatility as a writer, and to a political awakening later reported in his memoirs, but also to the flexibility of a 'discourse' of vitality itself, whose combination of specificity and suggestion, authority and speculation, made it peculiarly adaptable to such new usages. It also suggests that vitality, in many forms and guises, was, from the late 1780s, very much a topical concern in public discourse, and this is suggested too by a sermon preached by Samuel Horsley, Bishop of St David's, for the Humane Society on 22 March 1789. A one-time member of the Royal Society, where he had been an ally of the same Charles Hutton who, in 1822, enlightened the *Newcastle Magazine* regarding Hunter's resuscitation experiment on Dodd, Horsley had already made his name denouncing Joseph Priestley's denial of the Trinity, and would go on to lead Tory high-church opposition to the French Revolution. In early 1789, however, 'the Principle of Vitality in Man' is Horsley's concern in a sermon (later published by the Humane Society) which illustrates both the contemporaneity of a debate around vitality, and the difficulties faced by those attempting to assimilate it into a religious discourse to which it posed fundamental challenges.

As though acknowledging this, Horsley begins his sermon by addressing the discursive difference between religion and natural philosophy, a difference which he both recognises and attempts to bridge. Calling a war between Faith and Reason 'unnatural', Horsley admits the existence of separate spheres of knowledge, asserting that one must not look to divine scripture for natural knowledge, but that 'Revelation and Science may receive mutual illustration from a comparison with each other'.[19] Such general assertions pave the way for the sermon's more particular concern: to translate the medico-physiological concept of 'vitality' into religious discourse and close down the potentially materialist concept of a life-principle which is not a manifestation of divine agency but a property of matter. To this end, Horsley, in answer to his own question regarding 'the true principle of Vitality in the Human Species', reiterates a biblical model of clay or dust animated by divine breath or spirit, depicting mankind as a compound of body and immaterial soul. The 'union of the immaterial soul with the body is the true principle of Vitality in the human species' and no philosophy 'is to be heard, that would teach the contrary'.[20]

Quite how far this language fails to address the questions around human animation investigated by Hunter and others becomes evident when Horsley turns to mechanical metaphors to elucidate his picture. His assertion of a body–soul duality enables him to present the body as a mechanism or clockwork, set into motion either by the soul or, in the case of resuscitation, the 'clumsy fist of a clown'.[21] Such language protects those practising resuscitation, under the Humane Society's auspices, from accusations of 'playing God', but its account of vitality hardly corresponds to that generated by vitalist physiologists, who are largely motivated by the inadequacy of mechanical models to explain bodily functions in animate matter. Limiting 'vitality' to only the difference between life and death, Horsley overlooks the other vital functions and activities – such as self-preservation, healing and regeneration – central to the physiologists' distinction between animate and inanimate matter, as well as to their abandonment of an earlier iatromechanism. By attributing vitality to the soul, Horsley risks implying that vegetables and plants, as also clearly alive, must have souls too – a possibility at least raised, in a very different vein and from a very different perspective, by Erasmus Darwin, as the next chapter shows. His attempt to correlate complex physiological abstractions with biblical formulations, for all his assertion of their potential mutual illumination, only illustrates the gulf between two very different discourses.

To be fair to Horsley, Hunter himself had used the word 'soul' (as had Whytt) in his attempt to describe the mysterious 'something' which generates the vital properties of animated matter. '[W]hat is simply mechanical, that is made of inert matter, must have, as it were, a soul to put and continue it in motion,' he asserts in his *Lectures on the Principles of Surgery*.[22] But Hunter turns to the term 'soul' in the absence of an alternative technical term and its use – as 'as it were' suggests – is clearly metaphorical. Where Hunter is attempting to adapt an established, religious and philosophical vocabulary to articulate new physiological possibilities, Horsley is busy translating that new language back to an older one. Nevertheless, Horsley's sermon demonstrates how far a new language of vitality and reanimation was both current and contested in the late 1780s, with a dispute over its exact meanings – and the implied nature and relations of mankind, God, matter and soul – being played out not only for the attendees at specialised anatomical lectures or readers of physiological textbooks, but to a wide, public audience such as might have heard, or read, Horsley's sermon. Horsley's very willingness to preach on the subject of vitality is itself evidence of how far a concept whose origins lie in obscure physiological theory has travelled, to attain a general cultural presence even if, in the process, its meaning has been altered and eroded. The precise details of Horsley's failure to fully understand 'vitality' as the physiologists did is less significant than the way his attempts to co-opt the word signal its transition from technical term to cultural buzzword – an indication, perhaps, of ready, if not fully accurate, cultural receptivity for a science of reanimation brought to prominence and made fashionable by the activities of the Humane Society.

One way of reading Horsley's sermon, delivered as it is on the very eve of events leading to the French Revolution, is as an attempt to shore up a post-Newtonian physico-theological consensus, a marriage of natural philosophy and established religion, which the increasing development of specialised forms of knowledge – such as Hunter's physiological expertise – threatened. John Hunter's brother William was happy to end his anatomy lectures with a conventional and deferential statement of the limits of man's knowledge of God's complex handiwork, but Priestley's denunciation of the Trinity, which had occupied Horsley's firefighting a few years earlier, demonstrated another way in which religious faith and scientific knowledge might be brought into more problematic relation. Horsley's sermon on vitality is just one instance of a moment in the late 1780s when vitality, resuscitation and reanimation – as ideas, theories, possibilities, speculation and even, sometimes, fact – circulated in discourses as distinct as theology, physiology and anatomy, and periodical and sentimental literature. Within a few years, a tussle over the meaning of a political rhetoric infused with imagery of reanimation and vitality was to refract and focus the larger political divisions which marked responses in Britain to the French Revolution. In the process, reanimation became increasingly represented not as enacted on the passive bodies of female, pathetic or otherwise sentimental objects, but as a necessary revival of the manly, patriotic heroism which, variously, was either looked to in order to forge a newly just political settlement or, alternately, to save the nation from imminent revolutionary peril.

Politics and physiology in the 1790s

If reanimation, vitality and resuscitation were in scientific and cultural vogue in the 1780s, in the often highly figurative verbal and visual political discourse of the 1790s bodies were again everywhere. In James Gillray's satirical prints, swollen or disfigured bodies signalled the grotesque and debased nature of contemporary politics, as in his print *An Excrescence; – a Fungus; – Alias – a Toadstool upon a Dung-Hill* (December 1791) in which Pitt, in his ninth year in office, is represented as a toadstool on a dunghill monstrously and seditiously taking on a physical resemblance to the king, or in a later image of a gigantic Pitt sitting on the Speaker's chair in Parliament, crushing the opposition with one outsize foot, whilst the other is kissed by George Canning (Figures 1 and 2). Anti-revolutionary rhetoric, whether from Pitt or Burke or Canning's *The Anti-Jacobin*, frequently equated the Jacobin 'threat' with the potential physical collapse of the invaded or diseased body of the nation: for the latter, the 'Jacobin faction' was a cancerous 'persevering, indefatigable, desparate' force insinuated into 'the bosom of our country', and Burke famously used a sustained medical metaphor to inform his diagnosis of the Jacobin 'eruptive diseases of the state' in his *Letter on a Regicide Peace*.[23] For such commentators, distorted, corrupted or sick bodies gave a monstrous and horribly palpable form to otherwise abstract fears for the national and political future, offering a means of representing the vast immaterial entities of contemporary politics – the people, the government, the nation – and bringing immediacy and focus to the endless words and constantly unfolding contests. The body is appealed to also by those on the opposite end of the political spectrum, so that Mary Wollstonecraft's analysis of Burke's excessive rhetoric becomes – in a description of the 'fumes' in his brain which have dispelled 'the sober suggestions of reason' – an attack on the discomposed bodily œconomy of the person who could have produced such disordered speech.[24] Wollstonecraft's implicit appeal to a properly ordered self, more completely under the control of a dominant reason, which emerges more strongly in her *A Vindication of the Rights of Woman*, perhaps suggests a Godwinian distrust of excessive passions and sentiments – which, in the future envisaged by Godwin's *Political Justice*, would fall away before the steady and inevitable progress of rationality. So persistent is the presence of the body in political discourses – even when it is distrusted as a source of contagion or disorder – that at moments it itself becomes the contested object of such debates, caught up in a tussle between, for instance, the ideal, regulated, hygienic and composed female bodies such as Wollstonecraft extols in *Rights of Woman*, and the mad, diseased, corrupted, debauched bodies depicted by Burke or Gillray. Such alternative bodies offer divergent visions of the future, accessed through alternative somatic presents: the discipline and exercise extolled by Wollstonecraft or Thelwall (whose *Peripatetic* champions the simple activity of walking as enabling the exercise of free speech and the 'enfranchisement of vision'), or the dissipation and stupor seen in Burke's more pessimistic analyses. Evoked as a means of representing, clarifying, dramatising political realities, the body becomes an explicit object of political rhetoric, its attentions addressed

An Excrescence;_ a Fungus;_alias_ a Toadstool *upon a* Dunghill.

Figure 1 James Gillray, *An Excrescence; – a Fungus; – Alias – a Toadstool upon a Dung-Hill*

too in Thelwall's determination to 'rouse' the interest and energies of his readers and listeners, its habits, weaknesses and strengths considered the best means of administering democratic reforms or diagnosing immanent national collapse.

As the rest of this chapter will demonstrate, such a language of bodies is, in the 1790s, one way in which opposing rhetorics of organisation, energy, regulation and government are codified and played out, for a newly enlarged reading public and in a linguistic and political struggle as significant within the larger political landscape as those over 'nature' or 'imagination', which

Figure 2   James Gillray, *The Giant-Factotum Amusing Himself*

have been analysed elsewhere, and to which it is connected.[25] In this struggle, ready rhetorical references to the body – its distemper, illness, infection
or health – participate in an ongoing conflict over the validity or applicability of such language, in an indication of contested political values just as

significant as the denunciation of the terminology of 'sans culottism' by Burke or its embrace by Thelwall.[26] As John Barrell, and Raymond Williams before him, have asserted, struggles over language are always indications of larger political contests, but whilst political differences are self-evidently at stake in such exchanges, reading the bodies of 1790s political language in the context of the contemporary physiological theories and debates addressed in the previous section suggests that this language of political rhetoric contains and refracts another, very particular tussle going on over the nature of the body itself: what it is, how it is animated, how to cure it when it is diseased, how, even, to return it to life from near-deathly slumbers.[27] Depending on how answers to such questions might be anticipated, quite differing metaphors of the body politic, used to intervene in political debates, could be offered, each with their own account of a 'healthy' or proper relation of the people to government and power. In fact, the radically divergent versions of bodily health and function which are offered within political discourse at this time signal just how far the body, and alternative theories of physiological œconomy, were being reimagined or contested. Like the electrically charged vessels of a contemporary natural philosophy experiment, or like the mixed passions described by Hume, which gain strength through conjunction with each other, these two very different debates – on politics, on the body – can appear mutually attracted and fused, as though the startling, horrifying or beguiling prospects on the immediate political horizon could only be anticipated or described through a language already charged with the controversy of another, pre-existing debate.

It is not only live bodies, but dead, near-dead or ghostly reanimated ones which recur through the political discourse of the 1790s, and these figures are especially marked in the writings of Edmund Burke and the responses to Burke offered by radical lecturer and journalist John Thelwall – the same Thelwall who in the previous decade, as we have seen, deployed the resuscitated body in a literary context as a focus for sentimental moralising. Burke's vision, in his first *Letter on a Regicide Peace*, of a 'vast, tremendous, unformed spectre' which had risen 'out of the tomb of the murdered Monarchy in France' is reworked in Thelwall's *The Rights of Nature*, a specific counter to Burke's text and one which reimagines the undead body not as the monster of political revolution but as the dangerously supine nation stupefied by the oppressions of unjust government, for whom only the exertions of a 'manly', energising reason offer an alternative to imminent death.[28] For both men, the dead or near-dead body evokes the urgency and importance of the current political moment, understood in hyperbolic terms as an apocalyptic limit point before which all known forms of government, justice and civilisation might calamitously collapse – or give way to democratic alternatives. The use, by both Burke and Thelwall, of the image of John Žižka, the Hussite general who commanded that a drum made from his skin should be used after his death to repel enemy attackers, is peculiarly fitting in this context, for it is precisely by beating on the figurative image of the dead body that Thelwall and Burke offer their respective alarms.[29] But however resonant an image the dead body is for these rhetoricians, in Thelwall at least that image is more than just a figure, for his political lectures and writings coincide with a retheorisation of the body and its powers of vitality even in the face of death, whose suggestions of reanimation when deployed in the context of a political struggle for the life and future of the nation surely demanded rhetorical exploitation. Thelwall's insistence on the immediate revival of political justice, whose life-spark has been kept alive – just – in citizens' breasts throughout the centuries, thus differs from Burke's dark warnings of the immanent overthrow of European civilisation not simply in the political principles which led them to such differing interpretations of the present moment, but also in an underlying understanding of the powers of life and energy in the body which for both men operates as the ultimate ground on which the truth of their words is written.

Thelwall is certainly ready enough to acknowledge his differences with Burke over 'nature', a term necessarily inflected within any debate about human or animal physiology. In *Rights of Nature*, he suggests that 'nature' for Burke includes everything marked by 'the hoar of ancient prejudice', whilst for him it names only what is 'fit and true' and tested by reason.[30] Thelwall's faith in, and commitment to, reason, a keystone of his political creed, is suggested too by his remarkable *Essay Towards a Definition of Animal Vitality* (1793), fruit of an interest in the Hunterian physiological debates of the early 1790s, Thelwall's engagement with which was eclipsed only by the extent of his ongoing involvement in radical politics. As evidence of his detailed knowledge of current, controversial, physiological theory, the *Essay* demands that Thelwall's use of bodily images in his political writings is regarded as more than empty speech, and provides an insight into how those images might be properly interpreted. Rejecting Hunter's location of a vital principle in the blood, the *Essay* suggests instead that the body's 'state of vitality' is triggered by an external cause, such as an 'electrical fluid' – a theory offered too in the 'Digression for Anatomists' in Thelwall's near-contemporary *Peripatetic* – and one which emphasises the body's relation to external animating forces at the expense of Hunter's postulation of its self-generating animating powers. For Thelwall, at the end of a closely argued pamphlet which devotes much of its energies to exposing the lacunae and illogicalities in Hunter's own writing, life is defined as 'that state of action (induced by specific stimuli upon matter specifically organized)'; its 'preliminaries' are 'specific organization' and 'specific stimulus'.[31] At a time when, as Philip Rehbock has shown, debate around vitality was divided between those who favoured an innate, or material, vital power, and those who adhered to a more traditional belief in vitality as superadded, Thelwall's own definition of life was a clever balancing act between, or reconciliation of, these opposed positions.[32] But its emphasis on the issue of 'specific stimulus', or the all-important, but unknown, energy, activity or causational power, was to find its parallels within Thelwall's political writing, as the next section will show.

The distinction drawn by Thelwall, between an already existing and singular 'vital principle', and a bodily organisation capable of being animated by some unknown external force, such as the mysterious electricity, became the basis for a more polarised and controversial public dispute between John Abernethy and William Lawrence in the second decade of the nineteenth century.[33] In the 1790s, however, whilst alternative positions are being rehearsed – as in Thelwall's paper – they have yet to harden into defined, opposed and publicly defended positions. Hunter's own views, however much they informed his years of research and experimentation, are rarely at the forefront of his published writing: they are most explicit only in the posthumously published *Treatise on the Blood* (1794) and references to the vital principle are given only in an aside in his paper on drowning. In the years following his death in 1793, Hunter's work evidently defined a compelling area of debate and controversy, but one whose pursuit could necessitate some difficult negotiations with existing institutional and discursive structures. Thelwall's essay on vitality was first presented to the Physical Society at Guy's Hospital, a group whose status was somewhere between that of an official institutional body and that of an informal club, and a subsequent paper of his, 'On the Origin of Sensation', delivered to the same audience, prompted outrage, heated debate and strongly felt division, which only ended with Thelwall's resignation. His later comment on this episode, identifying the controversy as fundamentally political rather than physiological, is given as the epigraph to this chapter. Unlike later flowerings of the vitality debate in the next century, such ideas in the early 1790s circulated, if at times hazardously, relatively free from institutional authorisation, at the borders of defined and acknowledged areas of knowledge, perilously on the edge between what might be publicly debated and what could not be countenanced – as Thelwall's Physical Society resignation, or Humphry Davy's suppression of a similar paper, an *Essay to Prove that the Thinking Powers Depend on the Organisation of the Body*, suggests.[34] Not yet the explicit object of a defined discipline – it is not until Lawrence that the term 'biology' is coined – questions about the nature and powers of life, the causes of vitality and the defining differences between life and death, exerted a powerful pull, flickering with possibility, promising final answers to questions fundamental to the understanding of the nature and identity of human life. Arguably, as the remainder of this chapter explores, such questions, and their associated language, circulating in unresolved, but powerful ways, resonant, suggestive and of the moment, were readily incorporated into other discourses both for their metaphorical potential and for the possible knowledge – a kind of imaginary resource of what might be postulated or speculated – which they appeared to offer.

All of this means that, at least on Thelwall's side, and even within the highly metaphorical political language of the time, references to the body cannot be understood as operating innocently, but as carrying with them at some level invocations or traces of a very specific debate. Alan Richardson has written of the need to uncover the 'ideological as well as scientific and cultural valences of key terms and ideals usually limited to a range of literary meanings', and such a decoding must involve an awareness of the contemporary controversies in which such terms participated.[35] Thelwall's *Essay*, with its recasting of a debate on a 'vital principle' into a search for an external cause of animal vitality, might have appeared to have replaced a focused concern on vitality as a specific identity with a looser, even vaguer, emphasis on the body's relationship with the external environment, but in doing so he helped to move a debate hitherto focused on internal physiology onto new ground. Thelwall defended the kind of conjectural move such as he made as necessary for opening new speculative vistas: the Introduction to his *Essay* asserts that researches on this topic unavoidably operate via conjecture, and the *Peripatetic*'s 'Digression for the Anatomists' offers a poetic exposition of human vitality produced by a 'pure electric fire' only 'as a conjecture sufficiently probable'.[36] Whatever the effects within a scientific discourse on the vitality of more open-ended postulations such as these, arguably one effect of Thelwall's willingness to deal in such conjecture and speculation was to enable a language of vitality to resonate more readily beyond the precise bounds of physiological debate – as the *Peripatetic*'s poem already indicates. Nicholas Roe has explored how echoes of the vitality debate recur in the poetry of Wordsworth and others in the late 1790s to articulate, for instance, the animating principle of 'something far more deeply interfused' in 'Tintern Abbey'.[37] Despite the renewed critical attention given to Thelwall in recent years, however, there has been little sustained attempt to explore the presence of vitalist physiological language in his political speech and writings in the 1790s, which critics have preferred to discuss under the headings of enthusiasm, rationalism or politico-sentimentality.[38] In his 'Digression for the Anatomists', Thelwall showed himself to be interested less in the exact nature of an unknown principle of vitality, more in exploring the expression and achievement of vitality in uniting, transforming and energising the body. His political writings, as the next section will argue, which repeatedly deploy a language of generation, energy, vitality and activity to give a very particular twist to his ostensible concern with reason, demonstrate a similar interest in the possibilities of reanimation both as the means by which to revive the slumbering body of the British nation, and as a way of understanding the intended operation of that writing itself on Thelwall's listeners and readers.

Thelwall and the politics of animation

Proclaimed by E. P. Thompson as one of the most 'considerable' theorists of radical reformist politics, Thelwall has become a prominent figure in recent critical accounts of the 1790s.[39] Aided by a recent anthology of Thelwall's political writings edited by Gregory Claeys, together with renewed interest in radical working-class political organisations such as the London Corresponding Society, such work has enabled Thelwall to be seen, alongside Tom Paine and William Godwin, as one of the foremost agitators for political reform.[40] A consensus on Thelwall's prominence, however, has not been matched by agreement on how his profuse writings, which include lectures, pamphlets, poems and journalism, might be categorised and understood. For Paul Keen, Thelwall's foremost commitment is to the 'rationalist creed of the Enlightenment', as he sought to use the power of reason to claim access to a public sphere, or a republic of letters, from which a lower-class identity would otherwise exclude him. In this account, Thelwall's efforts in disseminating knowledge, dispelling ignorance and promoting political education enacted an ideal of communication and enlightenment which would enable him and others to storm 'the invisible walls of the republic of letters' and thereby bring about reform: for Keen, Thelwall offers a 'classic version of the reformist dream of the public sphere as an engine of non-violent social change'.[41] Jon Mee takes a different tack, assimilating Thelwall into an account of enthusiasm in the long eighteenth century, and seeing him as both offering political enthusiasm as a 'positive and liberating force' and as sharing larger cultural anxieties about the destabilising effects of enthusiasm as an unbridled energy.[42] Andrew McCann's reading of Thelwall's more literary writing, meanwhile, offers the term 'politico-sentimentality' to express his sense of how Thelwall appropriated 'affective sentimental discourse ... for radical political ends'.[43] Whilst all three critics share a recognition of the would-be performative nature of Thelwall's writings – that they were produced to further political agitation for reform – they differ in their characterisation of the nature of the activity which Thelwall's speeches and writings constituted: variously identified as rational enlightenment, enthusiasm or political sentimentality. Such differences are notable first because they repeat the ways in which political commentators of all kinds at this time themselves rehearsed varying formulations of, and attitudes to, political activity – Burke disparaged the 'restless agitating activity' and 'electric communication' of the Jacobins, and Joseph Ritson described as 'cant' the constant talk of 'force and energy of mind' from 'modern patriots and philosophers'.[44] But they also remind us how the nature of activity, its cause and the means of its stimulation was a central preoccupation of Thelwall's own writings (and of course, of the vitality debate to which, as we have seen, he contributed). Considered from this perspective, today's critical differences in interpretations of Thelwall could be seen as symptomatic of 1790s political discourses' own differences, in attitude and expression, regarding the activity which it understood as central to the struggles of the day. It also suggests that, when it comes to Thelwall, the critical imperative is less to categorise the activity which his writing both represents and enjoins, than to recognise the way activity itself, so necessary to the political aims he sought to achieve, as well as its possible causes, are themselves figured there.

Thelwall's much-debated fable of King Chaunticlere offers perhaps the most obvious, but also the most complex, indication of the writer's investment in questions of activity's cause, a question pursued, we might note, through a determined focus on the physical responses of two different bodies at moments of death. Published in amended form in Daniel Eaton's *Politics for the People*, the fable, subtitled 'The Fate of Tyranny', was offered by Thelwall during a Capel Court Society debate, and relates the beheading of the 'ermine spotted' tyrant cock, King Chaunticlere.[45] As John Barrell has noted, the tale recalls existing fabular representations of monarchy as game-cocks, and thus indulges in a favourite trick of Jacobin writing, offering a seditious depiction of regicide under a convenient and expeditious allegorical cover.[46] But it also offers a series of layered, alternative interpretations which makes the most obvious one look too superficial – or at least suggests that it is underlaid with a generative network of alternative hermeneutic possibilities. As Barrell explains, the tale can also be understood to reflect Thelwall's interest in what are termed the 'physical laws of the animal frame', and specifically the difference between involuntary, muscular motion, and voluntary, or willed, acts – a difference which played an important part in the vitality debate because involuntary activities were thought to point to the operation of some vital power. This interpretative thread is supported not only by the narrative attention given to the continued activity of Chaunticlere's headless body, as it runs senselessly around the farmyard after the cock's beheading, but also with a preceding anecdote, of a slave who, in the process of being burned as punishment for an attempted escape, raises an arm for self-protection against a blow intended to put him speedily out of the misery which an otherwise slow death would bring. As Barrell interprets it, this short narrative plays on the difference between muscular and voluntary motion to suggest that, just as the slave's involuntary raising of his arm is against his best interests, so too are the involuntary, unconsidered actions of other enslaved peoples (such as their automatic submission to authority) also counter-productive. What the distinction between voluntary and involuntary action points to, Barrell suggests, is the need for voluntary action: that is to say, determined, rational, conscious assertion of political liberties. In our involuntary responses, like those of the slave – unthinkingly trying to protect himself in a position where he is already doomed – we are little better than headless chickens. In this, Barrell's reading of Thelwall conforms with Keen's account of Thelwall the political rationalist: within the complexities of these dual narratives lies the injunction to self-determination of the political activist.

There remains however still another level at which these stories might be interpreted. Thelwall himself referred to Chaunticlere as a contribution to a debate on 'the love of life, the love of liberty, and the love of the fair sex', and considering it as such reveals another possible interpretation diametrically at odds with that retrieved by Barrell.[47] Thelwall's tales, of the slave and the game-cock, are offered, his preamble makes clear, to oppose a previous speaker's assertion that 'the love of life must certainly have the strongest influence on the actions of mankind', and to insist instead on the uselessness of life without liberty. His listeners should become 'acquainted with the real nature of that principle I am supporting' and 'learn to strike unanimously for liberty'.[48] What Thelwall's opponent might take to be evidence of an overriding love of life – the slave's raising of his arm against the cudgel, the cock's continued running after its beheading – could be seen as manifesting, at the fundamental level of involuntary muscular motion, a love of liberty without which life is not worth living. Viewed from this perspective, the tales offer not a parable of the need for rational re-education of an involuntary quietism, but evidence of a fundamental instinct, revealed in both slave and cock in the pressured moments before their deaths, for life *and* liberty. If, for Barrell, the fables can offer up a Thelwall insisting on the need for the involuntary instincts of our natures to be brought under the control of voluntary re-education, they might just as easily be seen to illuminate, and even celebrate, the existence of fundamental, instinctual, involuntary powers in our natures which pursue liberty before and beyond any subsequent acts of reasoning. If so, here is a Thelwall who not only heralds the rational education which will bring political freedoms, but one who locates the search for those freedoms deep within the 'physical laws of the animal frame' and its 'involuntary' principles.

The perhaps deliberate complexity and obscurity of these intertwined narratives should warn us against pushing these two alternative interpretations – liberty gained via reason, or via instinct – too much into opposition. Rather, we might note the striking way in which Thelwall's tales discover a coincidence or similitude between reason and instinct, both of which, according to which way the tales are read, could be understood as necessary in the 'strike for liberty' which Thelwall enjoins. Looked at this way, Thelwall's concern with the 'physical laws of the animal frame' appears invested less in distinguishing between voluntary and involuntary muscular action, than in offering a complex, sophisticated account of human nature and psychology for revolutionary times. What is most interesting, perhaps, is less the respective importance of reason and instinct, than the extent to which Thelwall harnesses the dual modes of the 'animal frame', instinctual and rational, to a teleological account of political liberation, which is thus made the proper outcome of every kind of human effort, willed or otherwise. It also suggests that if Thelwall offers a politics of rationality, such efforts might be understood – paradoxically, but powerfully – as releasing a love of and pursuit of liberty not acquired through reason, learning and enlightenment, but innate in human nature. Thelwall's politics of education, of undoing what is constrained and imprisoned in our natures, thus appears fuelled by a circular sense that the desire for liberty thereby sought is already located within our 'animal frames'. The efforts of reason are caused by an innate, vital sense of the desideratum of freedom.

This sense of an emancipating reason, which is not simply a rational force but a 'principle' of our natures, in itself offers a complex model of the kind of activity – at once both rational and instinctual – which might bring about revolutionary political change. This formulation also usefully constructs the activity of pursuing political freedoms as something for which a cause always already exists: the principle of liberty located within the human breast. Thelwall, the proponent of political change via rational education and enlightenment, as identified by Keen, is thus only part of the picture; by locating a constant 'principle' of the love of liberty, which operates even in moments of extreme stress and danger, Thelwall also asserts the existence of a force which will unwaveringly support and animate that quest for change. The problem, however, is that that force needs to be awoken and called forth, and thus, in extended and repeated figurations of the need for political engagement, the national body is again and again presented in his writing as slumbering, supine or near-dead in the face of decades of oppression. Not only the John Žižka image, already noted, but also that of a Sampson recumbent in Dalila's lap, and the stupor of the fallen angels in *Paradise Lost* are deployed to this end: the heterogeneous tools of historical myth, the Bible and literary epic subsumed into the creation of a new narrative of political urgency.[49] Thelwall's own role in such a story, like that of a 'resurrectionist' like Hunter, is to reanimate the nation with calls for renewed activity; not to *initiate* its political vitality, but to revive a spirit which has failed in the suffocating political climate of sustained injustice. Thelwall's reason, or 'energy' or 'activity' of mind, as it is variously termed, is thus not a cold Godwinian rationality or abstraction, but an almost organic force of political vitality, such that, just as the measures proposed by Hunter and Cullen will awaken the body's 'vital principle' and restore it to life, an inevitable 'quickening' of reason, produced by Thelwall's efforts, will propel the nation to a new life of justice and liberty.

Such a physiological perception of contemporary politics, which figures reason as an animated and animating force to return a supine nation to flourishing health, is often explicitly manifested in Thelwallian rhetoric. As we have seen, Thelwall's essay on vitality postulated the 'electrical fluid' as a possible source of the 'specific stimulus' which animates life – a suggestion pursued also in John Hunter's experiments on the torpedo or electric fish, and which appeared proved in further electrical experiments on animals published by Galvani in 1791.[50] Electricity, a favourite Burkean metaphor for an uncontrolled force of dissipation and social corruption, is in Thelwall by contrast repeatedly used to suggest the animating force of his rhetoric. A letter to his wife reporting the effects of his lectures described how every 'sentence darted from breast to breast with electric contagion', and electricity appears in the lectures themselves as an example of the galvanising force Thelwall would invoke:

> In principle, then, not in violence I would have you go forward. In active exertion of mind, not in tumult, I would have you advance. Now is the time to cherish a glowing energy that may rouse into action every nerve and faculty of mind, and fly from breast to breast like that electric principle which is perhaps the soul of the physical universe, till the whole mass is quickened, illuminated and informed.[51]

The energy of mental exertion, operating in the political arena, acts as a 'specific stimulus' to 'rouse' both individuals and the 'whole mass' of the people. Elsewhere, reason is, like a vital spark, a 'divine principle' which charges passive matter with energised life, or juries, with their judicious enactments of disinterested mind, are proclaimed, in the published version of the speech Thelwall wanted to offer for his defence in his own trial for treason, the 'soul' or 'vital spirit' of British liberty.[52]

This insistence on energised reason as the animating force of the body of the nation at times mounts to the rhapsodic expression of the universe itself as an animated, harmonious, entity, inhabited by some living power. Such expressions depict the universe as 'one continuous system of animated being' or speculate on the 'immortal' energy of actions which continues to resonate even after the death of their perpetrator.[53] At its most effusive, as perhaps in the first of his *Political Lectures*, Thelwall's rhapsodic depiction of a universe united by the 'nerves' of fellow-feeling, or his call for his 'bosom' to be animated with 'the enthusiastic love' of truth and justice, he recalls the Shaftesbury of *Characteristicks*, and can sound like the exponent of enthusiasm identified by Mee.[54] But the rhetorical swirls and flourishes of this lecture were surely designed to confound the government spy whose presence in an adjoining room Thelwall delightedly exposes, and in any case Thelwall's words are fundamentally rooted in a perceived animating force very different from enthusiasm. Whilst some passages in his writings are more effusive than others – a technique he himself acknowledged – taken as a whole, Thelwall's writing exhibits extended and closely worked argument, careful construction and exhortation to a reason quite distinct from enthusiasm. In his writings as a whole, and especially in two pamphlets of 1796, where an explicit engagement with Burke hardens Thelwall's language to produce a sharper and clearer account of his political beliefs, Thelwall's repeated insistence on the animating force of reason operates not as rhetorical vacuity or emptily heated language, but as the figurative expression of a sustained political philosophy.

It is Burke's attack on public opinion in his *Letters on the Prospect of a Regicide Peace* (1796) which elicits the first of these responses and reveals the specificity of Thelwall's deployment of a political language of animation.

In the first letter, Burke suggests that, whilst public opinion can be a significant national strength, the public mind can be mistaken, may need to be
'managed', and in any case, 'the British publick' can probably be computed
to consist of approximately 400,000 persons.[55] His account of the wars
fought by Britain under William of Orange, in which mental resolve was
the nation's greatest strength, shows that he can envisage positive examples of the 'energy' of the public mind. But in that historical anecdote, it is
William who 'infuses' the spirit of the nation to create 'one body, informed
by one soul': an exemplary instance of monarchical leadership of a nation
needing to be shaped.[56] Where the energy of the public mind operates in
a less controlled fashion, Burke's language becomes much more pejorative,
to figure such activity as a dubious, unregulated, damaging force. His sense
that 'strong passions' awaken the faculties of the people to use 'all their
native energies', or that ambassadors to France will become 'conductors of
contagion' or 'electricity', repeats a Thelwallian vocabulary, not to recognise
such forces as rightfully participating in Britain's political life, but tinged
with anxiety at the challenge which containing such forces presents.[57] Burke
prefers to retreat from the difficult specificities represented by individuals
and their opinions, formulating an account of 'the body politick of France',
for instance, as constituted not by the revolutionaries, but 'the majesty
of it's throne ... the dignity of it's nobility ... the honour of it's gentry ...
[and] the sanctity of it's clergy [*sic*]': an identification of the nation with its
abstract social and political institutions directly at odds with Thelwall's own
assertion that Britain's body politic consists of the aggregate of its people.[58]
Equally, the 'sovereign reason' celebrated by Burke, as Thelwall adroitly
demonstrates, is not the accumulated mass of human wisdom, but a mythic
entity located somewhere beyond existing human thinking and institutions.[59] Where Thelwall understands the accumulated energies of its people
as a vital national force, Burke retreats from such potentially divisive forces
into more manageable abstractions.

Thelwall offered his response to Burke in *The Rights of Nature, against the
Usurpations of Establishments* (1796), where he suggested that Burke's desire to
suppress public opinion was worse than war on liberty or property; indeed,
it is a 'war' on the 'sentiments and feelings of the subject multitude'.[60] This
association of Burke with death and destruction continues with an extended
figuration of Burke as a mad physician, attacking low fevers with amputation
and a cautery iron, for whom Jacobins are 'to be cut and burnt away, like
warts, from the eruptive body'.[61] Such surgical imagery has already been used
by Burke in his call for action to purge the national body of the sickness of
Jacobinism, and Thelwall readily accepts a rhetoric of the diseased political
body (wittily identifying Burke himself as a symptom) only to offer a medicine
quite at odds with Burke's. Political establishments do become corrupted over
time, he admits, but the disease of the state 'carries, in its own enormity, the
seeds of cure'.[62] The human tendency to communicate, the collective nature

of society, even the combinations of the workplace, all promote the circulation and acquisition of political knowledge, and hence reason and liberty. These myriad workings of human reason operate, Thelwall suggests, as the life-principle of the national body, the guarantee of its vitality. Reason can thus be celebrated as the 'phoenix mind' whose restorative capacities regenerate the nation at a point of near-death.[63] Like the hidden workings of vital forces, which, as we saw in Chapter 3, for Adam Smith had guaranteed the health of the national economy, reason's operations are naturally diffused to the advantage of the whole. As he elaborates in *Sober Reflections*, another response to Burke, such is reason's healing force that it causes the 'parts' of a potentially divided country to 'firmly and peacefully cohere' to 'one common centre of truth'; thus 'the planet shall roll on ... and the meteors of delusion shall burst and expire'.[64] Thelwall's reworking of a language of disease and corruption inherited from Burke thus extends, at its most developed, to conceive of reason as a benevolent power of the universe, an inevitable, insuppressible force of nature guaranteeing truth and peace even at a cosmic level.

But even if Thelwall's rhetoric reaches for the stars at such moments, unlike Burke he still conceives of reason as a human power, even if it is a 'divine' one too. Unlike Burke, who differentiates between a dangerously unstable public opinion and an abstract and unlocated 'Sovereign Reason', Thelwall is happy to equate the two: 'the most confident deduction of the most cultivated reason is but "an opinion" still'.[65] For Thelwall, public opinion operates as another term for reason: opinion is the accumulated expression and manifestation of individual acts of reasoning. Such is the power of public opinion, Thelwall insists, that it has an influence on government even when that public is not enfranchised with the vote. In his most extended figuration of the influence of public opinion, Thelwall borrows from Burke's favourite gothic mode to depict opinion disturbing the sleep of Pitt, where

> it requires no second-sight to perceive, that it haunts his imagination, and disturbs his slumbers. There, in prophetic visions, it fortels [*sic*] the sad catastrophe of his vision, and points out, in the *continuance* of this war, the means of British renovation; the approaching failure of the funding system; the demurs of money-lenders, and the prudent desertion of those 'life and fortune men' who ... finding the bankrupt state no longer competent to support, at once, the burthen of public credit, and the enormous prodigality of the present system, will be obliged to *abandon the borough-mongers to preserve their property*.[66]

Excluded from proper participation in the political destiny of the country, public opinion finds another means to intrude itself, in a ghostly return of the repressed. Thelwall's ghost recalls those used by Burke and others to depict the Jacobin forces which, for Burke, rise 'out of the tomb of the murdered Monarchy in France', or, in Gillray's prints, personify the fears, anxieties and madnesses incubated by the political events of the day.[67] In such contexts, the spectral embodies a feared future: Burke's apocalyptic vision of the collapse of European civilisation. But Thelwall's ghost has a rather different status, not a nightmare or malevolent force, but a necessary prophecy or timely warning. It signals something living on despite every attempt to exclude it, whose persistent half-life implies its innate power and presages its potential reanimation. This vision of a persistently vital public opinion is consonant with that given in his accounts of British history elsewhere, where public opinion's gradual growth over time and despite attempts at suppression is traced; Thelwall's image offers the ghost of public opinion as one which should properly be returned to life, in a reanimation perhaps effected in part through the force of Thelwall's own interventions, as he calls on his readers to 'let not only the nocturnal phantom, but the living body of your complaints appear before your oppressors'. Such a happy reanimation is anticipated too in a footnote, where the 'wandering ghost of popular discontent', buried through the efforts of the two gagging acts of 1795, is reported to be 'in hopes of a joyful resurrection'.[68] If Burke calls on ghosts to keep terrors 'artfully alive', as Thelwall accuses him, for Thelwall they reiterate his constant theme of necessary revival.[69] Where Burke offers, in Thelwall's terms, 'narcotics and soporifics' to 'charm us to sleep', Thelwall shows how even the deathly terrors of Gothicism can be reworked to further a politics of animation: the powers of reason implicit in the public opinion which haunts Pitt's sleep have a vital force stronger even than the death which Thelwall's opponents would wish for them.[70]

Language, vitality and the public sphere

Thelwall's reworking of Burkean Gothicism, to transform midnight horrors into welcome portents of future reforms, is of a piece with his usual strategy of response to Burke. Given Burke's often excessive imagery, for which he was widely ridiculed and criticised, Thelwall on the one hand adds his own voice to criticism of Burke's style, and on the other uses the images, metaphors and language bequeathed by Burke to inform his own political exposition. Nowhere is this more clear than in *Rights of Nature*, where within the first few pages Thelwall both attacks a Burkean style seen as complex and obfuscatory, disordered and unregulated, *and* borrows from Burke's references to John Žižka to offer his own stylistic dramas, opening his pamphlet with the extended image of the country slumped near death. At a time when, as John Barrell and others have shown, Burke's inclusion of 'imaginative' language within political discourse laid him open to derision, as well as accusations of madness (in which Thelwall joined), Thelwall's dual response to Burke, of indebtedness and attack, already suggests something of the complexity of political rhetoric in the 1790s.[71] As the attacks on Burke imply, this was a time when conventions regarding proper language use in political debates both existed and were being undermined – not only by Burke, but also by reformist or Jacobin writers who, without educational access to what Olivia Smith has called 'refined' language, forged their own literary or linguistic strategies as part of their claim to rightful political participation.[72] Allegory, as Michael Scrivener has demonstrated, was one means used to such ends, but so were fables, poems and songs, as well as the 'plain' language of Paine, or the codewords and the heavy ironies of Eaton's *Politics for the People*.[73] Given such creative diversity, perhaps the most dominant characteristic of radical writing in this period was its novelty: the 'new kinds of reading and writing' generated by Jacobin literary culture explored by Scrivener.[74]

Language use is a recurring topic for comment in Thelwall's own political writings. His attacks on 'innuendo', particularly in the context of treason trials which seek to twist words to manufacture damning evidence, are particularly noticeable, but other 'false' forms of language use are equally denigrated.[75] Burke's use of the letter form, for instance, in his *Letter to a Noble Lord*, is excoriated as a form of subterfuge, an untruthful formal device, and he is accused of mixing fact and fiction; Burke must also be in Thelwall's mind during an attack on those who find 'moonshine links of unconnected facts' in 'nature's loose analogies'.[76] A distinction between factual and figurative language, however, also operates in relation to his own language use. More than once in *The Tribune* he distinguishes between his use of 'facts' and his more 'rousing' language – the extended ornamental rhetorical passages which are often interspersed in his writings. When he is made weak through illness, he announces that facts alone will be offered, and a similar separation of rhetorical ornament from factual exposition operates in the two parts of *Rights of Nature*, where the second letter is as shorn of stylistic ornamentation as the first letter is fuelled by it.[77] This division, between truth and ornament, or fact and rhetoric, is perhaps what is meant by Smith, in her account of attempts to forge what she calls an 'intellectual vernacular', or a public political language for the uneducated, when she speaks of Thelwall and others 'being unable to align their political beliefs with their writings': in her analysis, the 'emotive heart and the political mind which Thelwall hoped to unite in his writings are kept apart by two styles of language, one for sentiment, another for thought'.[78]

Smith's statement implies that Thelwall's attempt to find a language style sufficiently expressive and effective for his purposes is ultimately unsuccessful, a failure she depicts as an inadequate integration of head and heart. The suggestion, to any student of Thelwall who has ploughed diligently through the many pages of his prolific and articulate writings, that he has somehow failed in his political self-expression, is a startling one, but of more relevance to the argument of this chapter is Smith's representation of that failure in terms of a dualistic economy of mind and body. Her analysis of 'two styles of language, one for sentiment, another for thought' is worth comparing to Thelwall's most explicit examination of the acts of reading and writing, offered in a response to Burke which, in its detailed pursuit of his conflicting meanings, is itself a conscientious act of textual criticism. A few pages into *Sober Reflections*, his response to Burke's *Letter to a Noble Lord*, Thelwall has already begun to unpick some of the 'absurdities' and contradictions of his opponent, but rather than dismiss his efforts, Thelwall welcomes them as contributions to a debate in which truth, through the work of reading, will ultimately triumph:

> [T]he very absurdities and sophisms of a vigorous mind ... from the energy with which they are expressed ... take fast hold upon the imagination, and compell [*sic*] the reflecting reader to give them that repeated revision which ... cannot fail of conducting the enquirer to principles of liberty and justice.[79]

Thelwall's is a fundamentally democratic model of reading, in which meaning and truth are not the possessions of the writer but of the 'reflecting reader'. In this textual economy, energetic writing, of whatever kind, 'cannot fail' to produce 'liberty and justice'. It is a statement which already says a lot about Thelwall's faith in his own practices of lecturing and journalism, and which contrasts with Burke's sense of individual subjects as all too often subject to delusion or ignorance, an assertion which justifies his sense of himself as sole possessor of a precariously communicated truth. But it is also an account which diffuses the distinction made by Smith, between mind and body, head and heart, in an account of workings of the larger force of 'energy': the energy of writing and of reading. A force larger than the individual acts of writing and reading, which furthers the ends of 'liberty and justice' even despite the nature of the written word, energy appears as a spirit which moves through writer, reader and text, to transform all into active participants in the arrival of political reform. Energetic writing and reflective reading are thus animated by a powerful active force – a vital spirit, perhaps – which can rework the dead letter of texts into a lively perception of 'liberty and justice'.

   This account of deriving truth through energetic reading may stem from Thelwall's own experience, attested elsewhere, of being converted to radical politics through an oppositional reading of Burke.[80] And just such an operation of energy, deriving 'truth' and 'justice' from Burkean 'absurdities', can be seen throughout *Sober Reflections*, in Thelwall's transformation of Burke's words from anti- to pro-reform. The operation of energy, in fact, bequeaths Thelwall a specific style and textual strategy, as he repeatedly mimics, inverts and reworks Burkean rhetoric to quite new ends. Whereas *Noble Lord* presents Burke as a self-appointed leader of the people, admonishing, guiding, protecting, Thelwall turns this round, praising Burke for knowing that the people needed to be roused from their slumber, but identifying his 'phrenzies' as the means by which this is achieved. He borrows Burke's tone to warn of 'inflammatory pamphlets and ferocious scurrilities in the daily prints', but then identifies Burke's own attack on property and aristocracy in *Noble Lord*'s broadside against the Duke of Bedford as the most dangerous of them all, whilst reiterating and extending the terms of that attack into an extended denunciation of property, power, class relations and social division.[81] Tongue-in-cheek, he praises Burke as the 'battle-axe of moral indignation' and for his 'mighty hatred', and only fears, as Burke himself might, that he has 'set a poison in circulation most dangerous to the health and existence of the social frame'.[82] Mischievously claiming Burkean chivalry for himself, he offers a mock adulation of Burke's talents – 'the splendid effusions of his inexhaustible fancy' – turning Burke's letter into an object of mock-awe, apparently submitting to a Burkean aesthetic response but in fact exposing it to derision. So completely does Thelwall adopt and rework Burke's language that even his own title, *Sober Reflections on the Seditious and Inflammatory Letter of The Right Hon. Edmund Burke*, begins to sound like a parody of the high-minded, thin-skinned communication of injury of *Noble Lord*. The irony and cleverness of Thelwall's own writing derive their energy so thoroughly from Burke's original text that he can no longer be understood to be offering 'Sober Reflections' but a deliberate, mischievous misreading of Burke, whose power originates in the strengths of the man he attacks.[83]

With such a textual strategy, such a mode of expounding his own argument and principles which depends so thoroughly on the satirical extension and inversion of his opponent's words, Thelwall here is perhaps at his furthest from the kind of Enlightenment rationalism with which some commentators have identified him. Reason is clearly at work throughout this text, but it is a reason which works through the specific textual operation of parody, mimicry, inversion and satire already described – not through abstracted systematic exposition of principle (though Thelwall is equally capable of that, as the second Letter of his *Rights of Nature* demonstrates). The reason which provides a route to political justice in this text is one which understands word play, mock effusions, rhetorical sleights of hand and the subtleties of tone in which apparent praise masks parodic attack; it is a reason which must be alert not only to the principles of each pamphleteer, but also to the language and discursive habits of each. Thelwall demands from his reader – or his listener, as many of his texts were originally delivered orally – not simply an ability to follow a reasoned argument, but a sensitivity to and comprehension of the larger spirit of its delivery, so that to read or hear Thelwall is not to arrive at a rational comprehension but to be awoken and energised. Thelwall's language use seeks, in short, not simply to communicate, clarify or persuade, as Tom Paine's might, to move or affect, as Burke's does, but to rouse its recipients to a heightened state of political being. In Thelwall's project of 'awakening' Britain, language is his key instrument of animation.

This account of political awakening through the operation of energy in language, by which truth, liberty and justice 'cannot fail' to appear, presents a very different model of Thelwall's relation to language and the 'public sphere' of politics than has been offered by some commentators. Paul Keen, Olivia Smith and Andrew McCann have all presented Thelwall as negotiating a problematic entry into a 'public sphere' or (for Keen) a 'republic of letters' from which those without extended education would be excluded. For each, the problem of accessing a public domain is understood as being addressed by Thelwall's yielding a newly fashioned tool, variously successful, be it a 'vernacular intellectual language' (Smith), rational education (Keen) or 'politico-sentimentality' (McCann). But the stress of these commentators on the difficulty of crossing the borders into the public sphere, even properly tooled up, only heightens the contrast with Thelwall's own repeated faith in public, political participation as a seemingly automatic and inevitable consequence of his or another's animating words – a faith which, at the textual level, constructs the corollary of a reader who need only read and reflect on words communicated with energy. In place of a definite boundary which needs to be deliberately and consciously stormed, Thelwall sketches inevitable participation, and the inevitable emergence of justice and liberty. Those whom he addresses already belong to the social whole, the 'harmonised mass' which makes an occasional appearance in his writings; at issue is not the claim to participation but the action which that participation demands, action understood not in terms of education, language or some other acquisition, but a proper comprehension of one's part in the social structure. Thelwall's own experience of just such a moment, when his political energies faltered following the death of his mother, offers a rare instance of political animation being illustrated via the example of himself. Rejecting the retrospection and superstition which has 'unnerved' man of the 'energies of his nature', he casts his melancholy from him, exchanging suffocating personal sentiments for a sense of his participation in the larger collectivity where his energies should more properly be expended:

> Be gone, ye idle, melancholy sensations; ye feelings that can produce no fruit ... [L]et me not, by unavailing regrets, and retrospective views, consume the energies to which I have no exclusive right ... For I am not a solitary individual. I stand not upon a world where I behold no inhabitant but myself. I am but a part – a little, little member of the great animal of human society – a palpilliary [sic] nerve upon one of the extremities! and I must do that duty to the whole, for which by my structure and organization I am adapted.[84]

Here the references to energy which are so frequent in Thelwall are brought into a more theorised context, as an excessive indulgence of personal sentiments is derided for diverting energy which properly belongs to the social whole. Read as a moment which dramatises the obstacles to political activity, it is personal feeling which is identified as inhibiting recognition of social cohesion, not a lack of reason or language – an analysis repeated in his contemporaneous *Poems Written In Close Confinement*.[85]

In contrast to the claims of today's commentators then, Thelwall himself experiences no difficulty in theorising membership of a public sphere, and is concerned only with properly animating the 'nerves' of the 'great animal of human society'. His rhetoric de-emphasises any notion of a separate target of a public sphere in favour of attention to a force or energy which motivates and activates what is then publicly manifested. He can thus be read not only as presenting animated human energies as already constituting the public domain of politics, but also as offering an alternative account of relations between private individuals and a public centre, in which the former enjoy an inevitable constitutive participation in the latter – if they could only see it. The man described by E. P. Thompson as the leading radical theoretician of the 1790s thus appears to have conceptualised political activity in ways which have less in common with Habermas-derived notions of a public sphere, and more in common, in a reliance on notions of inevitable participation in an organic social and political whole, on the organisational economies whose influence on Adam Smith's formulation of political economy was traced in the previous chapter. In fact, the suppression of private feeling for the public good, the downgrading of the individual self in favour of a rhapsodic recognition of a larger social or universal whole, recall Smith's discussion of Stoic philosophy in his *Theory of Moral Sentiments*, whose relation to his political economy I have discussed elsewhere.[86] One indication of Thelwall's conception of the relation of private subjects to a collective whole is given in his insistence on government's dependence, if it is to have any authority and power, on the approval of the governed: a model of automatic and democratic participation in which, far from being excluded, subjects control and oversee the centre.[87]

The language used by Thelwall in *The Tribune* to describe the structural cohesion which unites individual subject and socio-political whole looks, too, with its references to nerves and bodily extremities, to contemporary physiological accounts of the 'animal economy'. As we saw in Chapter 3, vitalist physiology, with its insistence on self-governing, cohesive, unified bodies, in which the smallest parts communicated with and acted on behalf of the larger whole, offered a new organisational metaphor which resonated beyond the confines of scientific discourse. In fact, eighteenth-century physiology's interest in organisation, as opposed to organicism, was one feature which differentiated it from later biological sciences. In their insistence on Thelwall as a standard-bearer for assaults on a public sphere or republic of letters from which he is excluded, critics have taken little account of an existing tradition of scientific education to which Thelwall, with his foray into the vitalist controversy, clearly has links. Neither Smith, with her bipartite model of refined or vulgar language, nor Keen, with his emphasis on the 'literature' of a republic of letters, comments on science as an alternative focus for education and knowledge in eighteenth-century Britain. Both William and John Hunter delivered well-attended anatomy lectures to students in London, and they were not the only sources of scientific instruction: many other teachers and lecturers offered similar courses. As Christopher Lawrence has noted, such an education, usually pursued for entry into surgical or medical professions, offered opportunities to dissenters whose religion barred them from university education, but it also offered access to disciplined modes of thought and language to those, like Thelwall, whose economic circumstances prevented them from pursuing more traditional educational routes.[88] Prominent examples of success stories for those taking this route included John Hunter himself, who became a leading teacher and researcher through an education acquired by working alongside his brother, initially as dissector and lecture assistant. Thelwall records in his essay on vitality that he did not attend Hunter's lectures, but he clearly knew their content in some detail, and appears to have had access to a student's notes. In the context of the availability of such highly developed and disciplined modes of scientific knowledge, notions of an exclusive 'republic of letters' run the risk of reiterating an exclusionary definition of knowledge and literate culture which such institutions as the Hunters' lectures and the scientific establishment challenged. Both Olivia Smith and Paul Keen, after all, quote from the records of Francis Place, as a representative member of the working class who sought to educate himself through reading, and whose list of 'useful books' includes amongst 'the histories of Greece and Rome', Hume, Smollett and Fielding, books 'relating to Science and the Arts' and 'some books on Anatomy and Surgery'.[89]

Thelwall's language of vitality, animation, force and energy certainly asks to be read in relation to this scientific context, whose concepts and terminology it repeats and reworks to new political ends, and in which he was immersed in the period immediately prior to his full engagement in radical activism. Indeed, it was only the strength of his commitment to political campaigning which caused him to drop his existing scientific interests. A 'strong' reading of the language of vitality in Thelwall, as this chapter has offered, traces the transference of vitality from scientific discourse into political rhetoric, where it theorises and upholds the possibility of political energy, animation and reform. It offers an account of the individual's relation to the social whole which repeats the vitalist emphasis on the automatic and constitutive relation of part to whole, and even extends to constructing an ethics in which the experiences of the individual diminish in importance compared to the good of the whole. It figures Thelwall's role, as writer and speaker, as the animating 'instrument', and his audience, the public body, as the object of his vitalising attentions – to be awoken to the energies of the mind, in direct contrast to the soporific rhetoric of Burke. So the final effect of the deployment of a language of vitality is its figuration of acts of language themselves as animating: language offers not simply 'facts', to use Thelwall's own opposition, but a means of 'rousing' and energising, working on his listeners and readers to perform the work of awakening them.

Whilst Thelwall's language invites these readings, however, it is also worth noting the ways in which it can escape precise analogies and comparisons with vitalist science. Exact parallels and similarities can slip away as everything – public opinion, reason, the mind – becomes subject to the animating energy which he repeatedly invokes. Thelwall's opposition to analogy no doubt mitigates against too determined correspondences between his political concerns and physiological theory: he has after all rejected following 'moonshine links of unconnected facts' in 'nature's loose analogies'.[90] But the use of a figurative language which operates more loosely than through exact correspondence speaks also to how Thelwall's rhetoric itself works: not via a series of confined, regulated and precise analogies but as itself a force of energy, referencing vitalist theories as a means of linguistically regenerating politics, invoking the vital life-spark which will animate political life as a way of giving life itself to that vital force. Freed from the need for formally policed similitude or equivalence, Thelwall's language can be truly generative, figurative without being tied to any precisely defined sense of what is being invoked. For all his rejection of analogies, Thelwall's rhetoric is of course riven with figuration and often operates at suggestive metaphorical levels – which is perhaps why the political sentiments examined in this chapter can be articulated so readily also in the more 'literary' forms of *The Peripatetic* or Thelwall's prison poems. Not a formally figurative language, but a creatively generative one, Thelwall's political rhetoric makes suggestive use of the linguistic possibilities of an adjacent discourse, so that his words, prospective rather than retrospective, engendering rather than confining, initiate the vital changes of which they already speak.

Part III
Vitalism, Animation, Culture

5
Animated Nature: Erasmus Darwin and the Poetry and Politics of Vital Matter, 1789–1803

In 1789, the following review of a recent work of poetry appeared in the *Analytical Review*, the progressive periodical issued by radical publisher Joseph Johnson:

> The poetry itself is of a very superior cast, and whether we consider the author's management of his subject, his delicacy of expression, or the sweetness of his numbers, we feel ourselves equally called upon to commend him. He introduces his various objects of description ... with so much versatility of genius, that we could not but admire the grace and ease, and the playfulness of fancy with which he conducts himself through this part of his business, perhaps the most difficult of all. His descriptions themselves are luminous as language selected with the finest taste can make them, and meet the eye with a boldness of projection unattainable by any hand but that of a master.

The reviewer goes on to praise the 'continued series of fictions' which enhance the 'beauty' of the poem, before, after a series of extended extracts from the work, concluding the review by hailing the author as a 'true poet'.[1] The reviewer was the poet William Cowper, the author was the provincial doctor and Lunar Society experimenter Erasmus Darwin, and the work itself was *The Loves of the Plants*, part of a larger work, *The Botanic Garden*, which would appear in complete form three years later.

Cowper's praise for Darwin's work continued in a later review in the same journal, in 1793, when the second and final part of *The Botanic Garden* appeared; not since Pope, he comments, has a poet seemed 'so privileged in the *verbosum curiosa felicitas*, or in delicacy and harmony of versification'.[2] Such was his admiration for his fellow poet that he was moved to express it in adulatory verses, which appeared in the third and subsequent editions of *The Botanic Garden*.[3] In his appreciation of Darwin, Cowper was far from alone. Horace Walpole commended 'the most delicious poem upon earth', he was admired by Percy Shelley and the young Coleridge, and other commentators rated Darwin above Pope and Milton.[4] Although the language of other reviews, both in 1789 and in 1792–93, is more measured than Cowper's, both *The Loves of the Plants* and *The Economy of Vegetation* were largely received as elegant, amusing and technically accomplished descriptive poetry which extended the pastoral genre in new directions. Strikingly, in view of the transformation of the public perception of Darwin, within a few years, into the controversial representative of radical science whom he appeared for some by 1794–95, both Tory-leaning publications, such as the *Critical Review*, and Whig journals like the *Monthly Review*, agreed on the value of his poetry and the worthiness of his poetico-philosophical project: the explication of botanical systems, and other natural philosophical phenomena, in verse form.[5] If the *Critical Review*'s author quibbles over the 'almost wholly erroneous' geology outlined in *The Economy of Vegetation* (where Darwin rejects a Mosaic history of the earth for one informed by the recent findings of contemporary geological science), the *Monthly Review* approvingly quotes Darwin's own apologetic defence of the bolder of his philosophical speculations: 'even *extravagant* theories, in those parts of philosophy where our knowledge is yet imperfect, have their use, as they encourage the execution of laborious experiments, or the investigation of ingenious deductions, to confirm or refute them'.[6] At the turn of the last decade of the eighteenth century, the example of Darwin at least would seem to suggest that poetry might occupy a cultural space almost as unriven by the divisions and controversies of party politics as the pastoral and mythic golden ages narrated within Darwin's verse itself.

By the time Darwin's last work, *The Temple of Nature*, appeared posthumously in 1803, however, both the political and the poetic climates had shifted dramatically. Not only was *The Temple of Nature* perceived, on both sides of the political spectrum, as a pro-revolutionary, anti-establishment work, but its poetic style, deemed so pleasing ten or 14 years earlier, was now judged to be repetitive, monotonous and harsh. Furthermore, the alliance of poetry and natural philosophy which Darwin was praised for pursuing with his earlier publications, was no longer to be tolerated. In part this might have been due to the nature of the scientific vision which *The Temple of Nature* offered – an account of the evolution of the world and human society over millennia from an original chaos, the gradual emergence and development of life forms into the numerous species of modern times, and the whole process nominally overseen by a Deity, but fuelled more directly by a pulsating, vital, life-force which regenerated both 'Monarchs and mushrooms' to new forms of organic life after their deaths. But reading the repeated condemnations of Darwin's poetical speculations in review after review, it is difficult not to feel that it is the affront of disciplinary transgression, as much as a challenge to religious orthodoxy, which is being registered in these reviews – or perhaps more accurately, that the latter is being received, and thereby somehow diminished and contained, as the former. Whereas a decade or so earlier a perceivedly apolitical poetry of nature could be aesthetically appraised in a comparatively and broadly unpartisan cultural space, in the early years of the nineteenth century the periodical press, of all political persuasions, is quick to recognise and condemn a poeticised exposition of a politics of nature which offered a fundamental challenge to social, political and religious establishments. It was particularly quick to counter it by invoking the bounds of proper literary practice which it deemed to have been transgressed. What has changed in the meantime is not only a more explicit unveiling of the nature of Darwin's scientific beliefs, there for those who could recognise them in his early poetry, but made unquestionably clear in his medical treatise *Zoonomia* (1794–96) as well as in his final poetic work, but also the political context in which such beliefs were received and appraised. The exposition of a vitalist, animated nature, whose love life in 1789 appears as harmlessly entertaining as the dressing-room pictures to which Darwin compared his verses, is by 1804 understood as dangerously seditious. The coterminous decline of Darwin's poetic reputation, from Cowper's praise to the condemnation he later received from Wordsworth, Coleridge, Southey and Keats, thus reflects not simply the change in poetic fashions with the rise of Romantic aesthetics, but also shows how the spurning by early nineteenth-century cultural establishments of the literary figure most prominently associated with vitalist science was never simply a matter of critical judgement. Within the history of the formation of poetic tastes in these crucially transformative years lies another story, of how poetry, however briefly, operated as a space for the 'reveries' of vitalist science, before such speculations were curtailed in the name of more harshly invoked disciplinary divisions, and for a poetry which, if and when it celebrated an animated and vital nature, did so in more abstract, less scientifically specific ways.

Darwin's vitalist poetics

Darwin's *Botanic Garden*, which has been described as the 'most highly praised book of 1792', consisted of two parts.[7] The second part, *The Loves of the Plants*, published first in 1789, offered a poetic exposition of the Linnaean sexual system of botanical classification, depicting for its readers in a succession of mini narratives the love lives of the animated and personified members of the vegetable kingdom. The first part of *The Botanic Garden*, *The Economy of Vegetation*, published in 1792 to allow for the completion of some of the experiments it reported, ranged widely through a generous survey of various natural philosophical phenomena, including on chemical, geological, biological and physiological topics, organised loosely around four cantos dealing successively with fire, earth, water and air. Both poems were supplemented with extensive notes, both footnotes (which could occupy up to half of each page) and lengthy additional notes at the end of each volume, some of which amount to short essays on their designated topics; the work as a whole thus constitutes a remarkable compendium of knowledge, theory and speculation on a breathtaking range of topics. A monument to the extent and breadth of Darwin's various intellectual interests, the poems thus ask to be read as more than mere entertaining verse – or, the poetic entertainment they hold out to the reader operates as a potential gateway to further reaches of scientific instruction. This was certainly Darwin's avowed aim in publishing the work, which he claimed would 'inlist [*sic*] the Imagination under the banner of Science' – a classic Enlightenment project of disseminating learning from the philosophers to the reading public, which was broadly hailed and welcomed by its initial reviewers.[8]

At the same time as Darwin was composing *The Botanic Garden* he was also at work on another publication, the medical treatise *Zoonomia; or, the Laws of Organic Life*, whose two volumes appeared in 1794 and 1796. According to Darwin's biographer, most of *The Loves of the Plants* was written between 1778 and 1784; at the beginning of this period, Darwin had already been at work for about eight years on the composition of what eventually became *Zoonomia*.[9] There are clear overlaps and correspondences between the two works: *The Botanic Garden*'s extensive scientific annotations, especially on physiological, biological or medical matters, address topics which will later be more fully elaborated in *Zoonomia*. Darwin clearly understood there to be a continuity between the scientific content of each work: arguments in *Zoonomia* are often cross-referenced back to discussions already published in the notation to the early poetry. The additional appearance of *Zoonomia* thus makes the already massive intellectual achievement of *The Botanic Garden* appear almost uncomprehendingly huge; it would also eventually make Darwin vulnerable to the accusations of 'system-making' which would be levelled at him when the tide of his popularity turned.

One key difference between the two works, however, for all their continuity, is that what had been confined to footnotes in the earlier work comes centre stage in the latter. Where the significance of *The Botanic Garden*'s assertions of the irritability or voluntary power of plants might not be clear even to the most assiduous footnote-reader, such central tenets of Darwin's vitalist account of animated nature are asserted in the opening pages of *Zoonomia*. There, his investigation of organic life begins with a denunciation of those, the 'idly ingenious', who have for too long sought to explain the 'laws of life' by 'those of mechanism and chemistry', and who have 'considered the body as a hydraulic machine, and the fluids as passing through a series of chemical changes, forgetting that animation was its essential characteristic'. Rather than looking beyond nature for explanatory models, enquirers should compare 'the properties belonging to animated nature with each other'; the analogies which exist in nature, a consequence of the 'great CREATOR' stamping 'a certain similitude on the features of nature, that demonstrates to us, that *the whole is one family of one parent*', would enable the laws of organic life to be identified, and a theory of nature based on animation's central properties to be founded.[10] Darwin then moves on to outline his own account of organic life's characteristic properties: the four kinds of observable motions attributable not to gravitation or 'chemistry' but to 'life'. These are: motions of irritability (motion in response to irritation caused by an external body); sensibility (in response to pleasure or pain); voluntary (in consequence of volition) and associability (prompted by association with other forms of motion).[11] The ultimate cause of such motions of life can only be hazily sketched, but Darwin suggests the existence of a 'living principle, or spirit of animation, which resides throughout the body, without being cognizable to our senses, except by its effects'. Whether such a principle is material or immaterial cannot ultimately be established, but some effort is spent debating the various possibilities. The 'electric aura' provides one possible speculative model, but the spirit of animation is considered to be more subtle than 'electric fluid', despite the evidence of recent experiments by Galvani and Volta showing electricity's power over bodily motion.[12] Gravity, after all, is an immaterial force which causes motion, and so too might the 'spirit of animation' be an 'immaterial agent supposed to exist in or with matter, but to be quite distinct from it'.[13] Then again, such a force may be 'matter of a finer kind', although a quote from St Paul reminds Darwin's readers that the ultimate cause of all motion is immaterial.

Despite such quibbles and the nod to religious orthodoxy, such speculations soon gained Darwin a reputation as a materialist – one whose account of life found little room for any explanatory causes other than material ones, and who, most importantly, rejected doctrines of the divine origin and maintenance of life. Darwin's labyrinthine lines of thought, which are suggestive rather than conclusive, are perhaps too indeterminate to demand such a label, but such (possibly deliberate) expositional subtleties were no doubt lost on many of those who associated him with materialism, a significant proportion of whom no doubt had no first-hand knowledge of his writings.[14] More pertinently, what Darwin outlines, in foregrounding an unknown 'spirit of animation' or 'living principle' (the terms are used interchangeably throughout *Zoonomia*) as the central animating faculty of life, is a vitalist theory: one which attempts to theorise life's operation on principles specific to organic matter, often by invoking an unknown and mysterious causal principle. The terms 'vitalist' and 'materialist' were to be opposed in the confrontation between Abernethy and Lawrence in the second decade of the nineteenth century, but, as Alan Richardson has pointed out, in the 1790s they were much more closely intertwined, and often (as Darwin's own case implies) difficult to distinguish.[15] The reception of Darwin's vitalist accounts of plant and animal life as 'materialist' itself demonstrates both the confusion of the terms, and the way accusations of materialism functioned not as precise descriptors of scientific theories, but as broad cudgels with which to beat back perceived threats to religious, political and social orthodoxies. At the same time, such a slippage appears to have enabled a vitalist language of animation, energy and life-force to somehow slip under the radar of cultural censorship and political controversy, which perhaps goes some way to explain its influence and longevity in political and literary contexts (as explored in the previous and next chapters) at a time when materialism itself was readily equated with subversion and heresy.

Read alongside, or in the wake of, *Zoonomia*, Darwin's poetry emerges less as it was initially received – as pastoral, descriptive nature poetry – than as engaged in the exposition of something much more scientifically specific: a sustained account of nature animated by vital forces and innate energies. In particular, the personification of vegetable life in *The Loves of the Plants* looks less like an amusing but frivolously indulged poetic conceit, and more like a deliberate, determined philosophical strategy. Most of *Zoonomia*, as a medical treatise, deals with human physiology and disease, but Darwin does consider the existence of the essential 'motions of matter' in plant life, and argues that each of the four motions can be identified in vegetables. Their irritability and sensitivity are demonstrated by the contraction of leaves and the opening and closing of petals; the existence of voluntary powers is assumed by their apparent suspension in the so-called 'sleep' of plants; and habitual movements of leaves or other parts are suggested as evidence for plants' powers of association.[16] It is Darwin's founding axiom of the 'analogy of nature' which underlies such assertions, and which also produces such passages as the following discussion of the similarity of various vegetable parts to what, for Darwin, are their animal equivalent:

> the roots of vegetables resemble the lacteal system of animals; ... the sap-vessels ... are analogous to the placental vessels of the foetus; the leaves of land plants resemble lungs, and those of acquatic plants the fills of fish; ... there are other systems of vessels resembling the vena portarum of quadrupeds, or the aorta of fish; ... the digestive power of vegetables is similar to that of animals converting the fluids, which they absorb, into sugar; ... their seeds resemble the eggs of animals, and their buds their viviparous offspring.[17]

Indeed, Darwin claims, 'individuals of the vegetable world may be considered as inferior or less perfect animals', a line of thought which leads him to assert that vegetable buds 'are affected with the passion of love', and even to ask whether 'vegetables have ideas of external things? ... [or] possess any organs of sense?'[18] With such questions, Darwin appears to have reached the end-point of what might be speculated within the pages of a scientific treatise, but such analogical possibilities evidently find new life within the pages of his poetry, where personified plant lives continue the speculations of his scientific work.

If analogy is central to Darwin's scientific method, it is also a key part of his understanding of the relationship between science and poetry. His advertisement to *The Loves of the Plants* speaks of leading the 'votaries' of the imagination 'from the looser analogies, which dress out the imagery of poetry, to the stricter ones, which form the ratiocination of philosophy'. The same sentiment recurs in *Zoonomia*, where the 'rational analogy' of science is contrasted with the 'licentious' analogy which ornaments wit and poetry.[19] Science and poetry are at once different, and on a continuum: the looser analogies of poetry can take over where the stricter discipline of science must stop. Darwin's methodological tenets here echo contemporary thinking in rhetoric and literary criticism, and in philosophy. George Campbell's *Philosophy of Rhetoric* (1776), for instance, notes that the resemblances between differences, such as are foregrounded by analogy, are striking and pleasurable in poetry, but can be unconvincing as part of a proof or argument in logic or philosophy.[20] Such commentary on the comparisons which analogy furnishes can be traced back to Addison's early eighteenth-century definition of wit as an unexpected resemblance between two ideas, and is echoed too in Thomas Reid's assertion that the pleasure and charm of poetry derive in part from the mental delight of pursuing them. But in philosophy, according to Reid, the 'way of analogy' can lead to 'error and delusion', potentially giving credence to what is spurious and questionable. The mental appeal of the surprising, unlikely or novel, so valuable in poetry, is dangerous where the logical caution of reason is expected.[21]

However they reflect contemporary critical sentiments, Darwin's poetic and philosophical methods constitute one area which comes under attack from the more hostile reviews which greeted *The Temple of Nature* in 1803–4. Even the Whiggish *Monthly Review* attacks what it sees as a 'loose and unqualified mode of guessing', and the *Edinburgh Review* identifies a 'presumptuous contempt, or perhaps a gross ignorance of the legitimate bounds of philosophical inquiry'.[22] In such responses, Darwin's distinction between the methods of poetry and philosophy appears forgotten, as the reviewers respond to a work of poetry as though it were science; their reassertion of the proper disciplinary methods appropriate to the genre of the work they see themselves to be reading is in effect a rejection of the Darwinian hybrid of philosophical poetry (a rejection often made explicit too in the more general judgements with which many reviews open). If Darwin seeks to continue his scientific speculations 'under the banner' of imagination, these reviews close the door to any such licentiousness, using accusations of the transgression of formal discursive rules as a means of suppressing a vision – or a 'religion' – of nature variously seen as 'bitter', 'perverse', 'degrading' and 'heathen'.[23] In the process, they are implicitly reinforcing an expectation that poetic writing is substantially and generically different from natural philosophy, and that literary or aesthetic productions occupy a different cultural space from the reasoned practices of science.

But even where Darwin's writing is explicitly assessed as poetry, his use of a particular version of analogy, personification, still draws fire from some quarters. In a digression on Darwin in a review of Chalmer's *English Poets* in the *Quarterly Review* in 1814, for instance, Robert Southey condemns a poetry which he sees as mechanical and laboured, which offers only a 'perpetual strain of cold and systematic personification' instead of life and 'grace'; the terms of his attack are repeated in a later review where the machinery of Darwin's poem – its use of personification – is considered to be 'but laboured allegory at best', though 'more frequently an allegorical riddle', and both 'preposterous' and 'wearying'.[24] Darwin is falling foul here of the emergent Romantic equation between poetry and the organic, articulated too in Wordsworth's 'Preface' to the *Lyrical Ballads*, where an insistence on poetry as using language of 'flesh and blood', and an attack on personification as a 'mechanical device of style', could be read as in part a rejection of Darwinian verse style from one who had been under what he later styled as the 'injurious influence' of Darwin's 'dazzling manner'.[25] Ironically, the power of animating language and poetry, so central to Romantic aesthetics, was in eighteenth-century criticism assigned to the very figure of personification which was later to be found so objectionable. For Goldsmith, for example, prosopopoeia is foremost amongst a range of figures used in poetry which 'serve to animate the whole, and distinguish the glowing effusions of real inspiration from the cold efforts of mere science'; it animates not only the verse itself, but also the represented external world, being 'a kind of magic, by which the poet gives life and motion to every inanimate part of nature'.[26] Since Dryden and Addison, indeed, personification's ability to furnish striking images had been regarded as central to its provision of the 'life and spirit' of poetry, and with increasing emphasis on and recognition of the creative powers of the imagination as the century progressed, personification became increasingly valued.[27]

As Anna Seward noted, in Darwin's poetry 'universal personification [is] the order of the Muse'.[28] Both the *Loves of the Plants* and the *Economy of Vegetation* are extensively populated with personifications, not only of individual flowers and plants, but also of such figures as the Goddess of Love, and fleetingly animated objects, such as a compass or thermometer. This might suggest that Darwin's poetry would best be understood – and defended from such attacks as Southey's – by reference to the high status placed on personification in the eighteenth-century critical canon. Certainly, Darwin's own comments on personification reflect the critical precepts of the early eighteenth century, but the deployment of the figure in both *The Botanic Garden* and *The Temple of Nature* differs significantly not only from those precepts but from the poetic practice of his contemporaries.

Personification is addressed in the first Interlude to the *Loves of the Plants*, an educative dialogue between Bookseller and Poet, where the Poet suggests that in contrast with prose, which employs many words signifying abstract ideas, poetry seeks to engage the senses, especially the sense of vision; personification, by making images of what is otherwise abstract, is an essential tool for this purpose. Personification is presented here as a device to transform the overly abstract language of prose and produce a more visual poetic language, the better to engage the reader's 'eye' and produce the images at which poetry aims. It is an account which recalls the critical thinking of early eighteenth-century attempts to regard poetry as a sister art to painting (*'ut pictura poesis'*), rather than the later association of personification with the creative imagination.

Not only his theory but also his practice of personification distinguishes Darwin from the concerns of much mid- to late century poetry. In Gray and Collins, for instance, personification is typically of abstract terms: Evening, Spring, Care, Contemplation; it is such abstractions, representing 'moral virtues, or inanimate beings', that prompts Goldsmith's praises.[29] Darwin, by contrast, in ascribing human passions to plants in *Loves of the Plants*, personifies not abstractions but natural objects, which are already living things. Equally, Darwin's personification takes place not in a poetry of enthusiasm and inspired imagination, but rather in a verse whose crafted and polished rhyming couplets constantly suggest control and containment, and in which the central conceit of humanised plants is a sign of the poet's detached wit, rather than imaginative inspiration. Darwin's personification does not, as it does for theorists Kames and Blair, derive from poetic passion, but rather offers a sustained joke for the amusement (and education) of its readers. Like a magic lantern show (another of his comparators for his verse), Darwin's poetry offers a series of images, entertaining, but no more transporting for the reader than they are the product of the poet's own transport: his verse, in his own words, is an 'unskilful exhibition in some village barn', to be regarded indulgently by the onlooker. In alternative formulations, the *Loves of the Plants* is a collection of portraits hung in a lady's dressing room, or a series of images 'wrapped upon rollers' and successively displayed, and the poet himself is simply a 'flower-painter'.[30] Such self-deprecating representations reinforce the Interlude's sense of the poet-as-picture-maker, but, by doing so, they also overshadow a more significant effect of personification in Darwin's poetry than the creation of image. Despite his presentation of the poet as visual artist, and despite the Interlude's account of personification as a device to make the abstract material and visible, personification's key function in Darwin's poetry differs from what he admits to here. The central purpose of personification throughout his verse is not to offer a visual image of an abstraction, but to further animate already visible, already living objects. Viewed from the perspective of his scientific interests, Darwin's practice of personification, in fact, arguably has more to do with extending qualities of life, emotion and consciousness to natural objects than it has with visualisation; his poetry does not simply draw pictures, but, in the words of the Proem, imagines and restores an 'original' animation. Viewed thus, Darwin's purpose, announced as 'restor[ing] plants to their original animality', can be seen to fulfil a scientific as much as a poetic agenda: to use poetry as a means of enabling a vitalist nature to be brought to imaginative life in a way not allowed within the stricter confines of scientific method.[31] Darwin's exposition, in *Zoonomia*, of a vegetable nature which, like all forms of life, is governed by a 'spirit of animation' and is vital, self-animating and self-directing, thus appears already fulfilled in the personifications of *The Botanic Garden*, in a way which *Zoonomia*, limited to 'rational' analogies, cannot fully establish. *Zoonomia's* speculation that the anthers and stigmas of plants are 'real animals' who are 'affected with the passion of love' cannot be convincingly demonstrated, and further speculations regarding the existence of a 'sensorium' in vegetable buds remain suggestive only of Darwin's vision of a more fully animated plant life. 'I think we may truly conclude', he speculates, that each bud is 'furnished with a common sensorium ... and that they must occasionally repeat those perceptions either in their dreams or waking hours, and consequently possess ideas of many of the properties of the external world, and of their own existence.'[32]

Such scientific fancies, preposterous in a work of natural philosophy, are given a life of their own in Darwin's vitalist verse. Where *Zoonomia* asks 'whether vegetables have ideas of external things', or 'whether vegetables possess any organs of sense', *Loves of the Plants* pictures the 'chaste Mimosa' or sensitive plant 'alive through all her form', recoiling at the clouds of a gathering storm, or 'sad Anemone' calling on the swallow, 'herald of summer', to chase away the intemperate climates of winter.[33] The questions which mark the furthest extent of *Zoonomia's* speculations on vegetable life are the occasion for *The Botanic Garden's* founding conceit, the 'original animation' of plant life, repeated and explored in instance after instance throughout the book's eight cantos. Refusing the 'strict' constraints of scientific analogy, *The Botanic Garden's* pleasurable and surprising connections between plant and animal life both further Darwin's vision of nature and prove amenable to established critical criteria of a poetry enlivened by novel and striking animation.

In his 'Preface' to the *Lyrical Ballads*, Wordsworth made the memorable distinction between the 'Man of Poetry' and the 'Man of Science', a division which can retrospectively be read as a recognition (or inauguration) of the emerging gulf between the 'two cultures' of modern times: science in its current, specialised sense, and artistic, including literary, activity.[34] In Wordsworth's account of their relationship, the man of poetry gives animating life and vital form to 'the objects of science', an act whose casting of poetry as, not a mere form of letters, but a cultural metadiscourse, also masks a dependence in poetry's relation to the science which it animates.[35] Darwin's use of poetry to, in Wordsworth's words, give the objects of science 'a form of flesh and blood', begs the question of whether, for him too, poetry's relation to science is similarly oppositional and hierarchical. On the one hand, Darwin's verse could simply be regarded as the straightforward poetic expression of his scientific speculations – a translation into poetry of a preconceived vitalist science of nature. On the other, it could be seen more purposively and strategically as enabling Darwin to voice something which could not be said, or imagined, in science. Here, Darwin's remarks on poetry as reverie, offered in the first Interlude of the *Loves of the Plants*, are especially useful. Darwin outlines a theory of poetry as reverie: poetry induces a train of ideas before the imagination which we follow as in a dream, released from the restraints of rational control or conscious will. Poetry and the imagination are thus equated – in anticipation of some later, Romantic theories – with fantasy, the unconscious and unrestraint, although Darwin's emphasis lies more on the potential for imaginatively recalling and reviewing ideas than on the creation of the new. *Zoonomia* also presents the imagination as a powerful, vivacious and pleasurable faculty. Ideas in the imagination, Darwin suggests, 'such as those of a person in ... a waking reverie ... are more vivid and distinct than those of memory' – in part because they are charged with pleasure or pain.[36] Such comments suggest that Darwin's poetry is not a space of freewheeling imaginative creation, but of the contemplative, pleasurable review of ideas vividly recalled, a meditation on received ideas peculiarly charged with pleasure and vivacity. Poetry offers neither a frivolous space of meaningless play, nor (as Wordsworth implies) a secondary articulation of previously established scientific discoveries, but a pleasurable replay and exploration of recalled ideas which might be either a preamble or supplement to science, or a fulfilling meditation in itself. In this formulation, poetry is neither an alternative to nor a servant of science: its loose analogies might precede, or follow, the strict connections of science, but they might equally find their own end in the pleasure they entail.

Darwin's self-proclaimed project of 'inlisting imagination under the banner of science' thus emerges as more than the simple educational project of using poetic entertainment to disseminate science. More complexly, imagination is enlisted as a means of extending the speculative possibilities of science and the mental acts appropriate to it, as well as dissolving the stricter disciplinary boundaries which define its activities. His vitalist poetics – the poetic reverie which sustains an animated vision of nature – is also equally different from the personification valued by eighteenth-century critics and the 'mechanism' identified by later commentators. For Addison, the personification which is so central to Darwin's reverie was valued because 'it has something in it like creation; it bestows a kind of existence and draws up to the reader's view several objects which are not to be found in being'.[37] For Darwin, personification, viewed in poetic reverie, is valued not because it presents what is not in existence, but because it allows a meditation on what existence could be conjectured to be. Not a sign of what, in the second half of the eighteenth century, is an increasingly licensed and valued imaginative creation, Darwin's personifications are a vehicle for a pleasurable vital poetics of imaginative recall and reverie. In turn, such a poetry of reverie implies a potentially interdependent, or continuous, relationship between Darwin's scientific speculations and his poetry, which a more narrow view of personification as simply a manifestation of the poet's imaginative creation would overlook. The developing theory of the creative activity of the imagination, whilst bestowing new value on poetry, also worked to divide it from those discourses, including science, which were perceived as products of more rational mental faculties. With a poetics of reverie, rather than of the creative imagination, Darwin keeps open the links between his science and his poetry which the contemporary poetics of imaginative creation might otherwise close down. The same links between his science and his poetry also redeem Darwin's personifications from Romantic accusations of an empty mechanism. Far from being so, they offer a pleasurable speculation, a dream-like contemplation, of a version of nature which, animated, vital, independent and self-directing, anticipates the conception of nature which Romantic poets themselves were to take to heart. Southey's characterisation of Darwin's 'cold and systematic personification' is all the more ironic because that personification aims to animate a version of nature which Southey's Romantic peers would find congenial.

Perhaps most prominently amongst those peers, Coleridge, as he retreated from his youthful radicalism, would find, in the early years of the nineteenth century, that a recognition of nature's vital, self-animating powers could even be reconciled with religious faith, a position which enabled him to support the self-styled vitalist Abernethy against Lawrence's apparent materialism in the vitalist controversy of 1817.[38] In the fevered climate of the 1790s, however, during war with revolutionary France and fears of insurrection at home, the same account of nature, as vitally animated, self-directing and self-evolving, would be regarded by loyalist propagandists as amongst the most dangerous doctrines of radical experimental philosophy. As is shown later in this chapter, however, Darwin's work could also be co-opted by writers and publishers – such as Thelwall and Daniel Isaac Eaton – agitating for political reform. But to trace responses to Darwin's writing in the 1790s is not simply to illuminate the familiar opposition between pro- and counter-revolutionary propagandists, but, more precisely, to demonstrate how their political differences were supplemented with quite divergent visions of the place and purpose of poetry in national political debate. Because of his near-unique position as both poet and scientist, Darwin's inclusion in the controversy over radical natural philosophy in the 1790s thus also carried consequences for the fate of poetry in the new cultural settlements of the day. The same rejection of the Darwinian hybrid, of philosophical or scientific poetry, as would later be formally enacted in the reviewers' response to his last work, *The Temple of Nature* – where it appears as a formal aesthetic preference for a 'purer' poetry unsullied by cross-disciplinary ambition – can thus be seen as the consolidation of a rejection of Darwin's writings stemming not from aesthetic principle but from the culture wars and political battles of the previous decade. In turn, that same rejection signals the apparent triumph (in the would-be high-culture circles of the periodicals, at least) of a Tory and loyalist version of literary culture over the more democratic and heterogeneous models deployed in the radical publications of the previous decade.

Vitalism and poetic propaganda

In the propaganda wars of the 1790s, Darwin's vitalist science of nature achieved a renown, and a notoriety, it might otherwise have foregone, through being linked in the public mind with dangerously radical science. George Canning's *The Loves of the Triangles*, published in three instalments in the government-sponsored weekly periodical *The Anti-Jacobin, or Weekly Examiner*, between April and June 1798, is the most prominent of the satirical attacks on Darwin's science produced in the propaganda wars of the mid-1790s.[39] A parodic response to *The Loves of the Plants*, which aimed, along with a promised sequel, *The Algebraic Garden*, to 'enlist Imagination under the banners of Geometry', Canning's poem took aim at both the form and content of Darwin's verse, mimicking the mix of philosophical, scientific and prosaic topics, whilst making that mix look disordered and uncontrolled, and using the same extensive footnotes as Darwin himself employed to attack the detail of his philosophical theories. Canning's regular lampoons of liberal and radical writers were an influential part of the journal's attack on their ideas and politics, and the effectiveness of his travestying of Darwin's verse was such that it was still noted in reviews of the posthumously published *Temple of Nature* (1803) five or six years later. Indeed one reviewer attempted to extend the joke – and the attack – by tracing similarities between Darwin's final work and Canning's parody, suggesting that the poet was dunce enough to plagiarise a text in which he was himself ridiculed.[40] As a key part of propagandist response to Darwin, *The Loves of the Triangles* demands detailed attention, but the battlelines which lie beneath its polished witticisms are perhaps most usefully approached via the earlier, less honed and thus in some ways more overt broadside embodied in an anonymous poem, *The Golden Age*, which had appeared four years earlier.

Published in 1794 by the loyalist press of Francis and Charles Rivington, who had a year previously founded the periodical the *British Critic* to uphold the values of the established church and the Tory party, *The Golden Age*'s hostility to Darwin marks out the kinds of attack which, made repeatedly and often in stylised or coded form, would reappear in subsequent satirical works, both verbal and graphic, over the next decade. Ostensibly voiced by Darwin himself (who felt obliged to publish a notice in his local newspaper disowning authorship), the poem is presented as praising the work of Darwin's fellow scientist Thomas Beddoes. Recently refused the Regius Chair of Chemistry at Oxford due to his enthusiastic welcome of the French Revolution, Beddoes went on to found the Pneumatic Institute at Bristol where Humphry Davy would be employed, and where Coleridge and others would famously experiment with nitrous oxide. Beddoes' optimistic vision of immediate and palpable improvements in public health and living conditions, stemming directly from imminent scientific and medical advances and leading in turn to political reform, offered both a succinct expression of the social and political benefits sought by other radical philosophers (such as Priestley and Bentham), and fertile ground for attack from those either less convinced of the need for such improvements or hostile to the means by which they were hoped to be brought about.[41] *The Golden Age* takes its cue from one of the more unlikely expressions of Beddoes' faith in the transformative effects of science, his anticipation that animal products might one day be grown on plants, and that 'our Woods and Hedges' might thereby be taught 'to supply us with Butter and Tallow'. All too readily implying that scientific 'improvement' in fact constitutes a perversion or reversal of nature's norms, his suggestion prompts an extended attack on visions of scientific progress, which are made to seem fantastical and deluded, socially dangerous and, by extension, irreligious, anti-monarchist and revolutionary. Scientific 'meddling' in organic nature is associated with a desire for disruptive change in other areas of 'nature' – social hierarchy, political structures, established religion – and thereby experimental philosophy, as well as being debased on its own terms, is made interchangeable with seditious, revolutionary sentiments. Beddoes' own close association between his science and politics can thus be turned against him, as in a footnote which quotes his desire not simply to see valuable social change resulting from scientific advances, but his faith that 'Liberty and Genius' will in turn produce new blessings in the production of further bounties of nature.[42]

Whilst the most obvious target of *The Golden Age* is Beddoes himself, Darwin is implicated in the attack, both by association and by specific barbs aimed more directly at himself. Beddoes' faith in the vegetable generation of animal products is supported by reference to the botanists' claims of plant/animal similarity (a central tenet, as we have seen, of *The Loves of the Plants* and *Zoonomia*), and the poem finds room for a satirical treatment of Darwinian sensitive plants:

> See plants, susceptible of joy and woe,
> Feel all we feel, and know whate'er we know!
> View them like us inclin'd to watch or sleep,

Like us to smile, and, ah! like us to weep!
Like us behold them glow with warm desire,
And catch from Beauty's glance celestial fire!
Then, oh! ye fair, if through the shady grove
Musing on absent Lovers you should rove,
And there with tempting step all heedless brush
Too near some wanton metamorphos'd Bush,
Or only hear perchance the western breeze
Steal murmuring through the animated Trees,
Beware, beware, lest to your cost you find
The Bushes dangerous, dangerous too the Wind,
Lest, ah! too late with shame and grief you feel
What your fictitious Pads would ill conceal![43]

The animated nature which benignly pursues its loves in Darwin's *Botanic Garden* is here transformed to become both ridiculous and predatory, as the wandering virgin, molested by amorous bush or wind, unfortunately discovers. The scene's replaying of the seduction narrative – albeit in comic vein – equates the vital energies of Darwinian nature with social jeopardy and sexual transgression, and with the dangerous unleashing of desires more properly contained in established bounds. Darwin's readiness to portray a coy and indulgent pursuit of sexual pleasure in his poems would be targeted too in other attacks – in *The Loves of the Triangles*, where his celebration of mythic female beauty is bathetically reduced to the eccentric poet's voyeristic spying on young ladies taking the air in a London park, or in the association made by Canning's later *New Morality* between Darwinian sensibility and sexual, and hence political, susceptibility. Given the importance of sexual reproduction in Darwin's account of a self-regenerative nature (and its superiority over asexual reproduction would be a significant theme in *Temple of Nature*), such attacks, whilst sullying Darwin's reputation by figuring him as a purveyor of cheap sexual titillation, also work to resist his insistence on nature's innate energies and powers, by reading such vitalist tenets not as cornerstones of a developed system of nature, but as personal expressions of (at best, improper, and at worst, revolutionary) authorial sexual licentiousness.

Aside from the amorous bush, and references to oxygen – the 'vital air' – which would be developed in *Loves of the Triangles*, the significance of *The Golden Age* in terms of the developing attack on Darwin lies mainly in the poem's movement from the specifics of natural philosophical beliefs to an uncovering of their ideological import. The poem's broadening, in its final pages, from the satirical representation of Beddoes' and Darwin's work, to their putative seditious meanings – attacking rank and distinction, established social order, religion and monarchy – offers a structure which is repeated in both *The Loves of the Triangles* and, more obviously, in *New Morality*. Such a move prepares the ground for more broad-based attacks on Darwin where the specifics of his science are rarely addressed, but where a 'New Philosophy' (*New Morality*) or 'New Principles' (*Loves of the Triangles*) can instead be invoked without holding the progress of the poem up by detailed denunciation of scientific principles. This no doubt goes some way to explain the repeated presence of Darwin in a range of verbal and visual attacks on the perceived embodiments of radical and liberal sentiment in the next decade. He is included, for instance alongside Godwin, Priestley, Horne Tooke, Thelwall, Price and others, in Gillray's *New Morality* engraving (Figure 3), despite not being named by Canning in the earlier poem of the same name, and appears in similar company in the 'Dunghill of Republican Horse Turds', in Gillray's 1800 print, *The Apples and the Horse-Turds* (Figure 4); his writings, or 'Darwin's topsy-turvy Plants and Animals' Destruction', lie amongst the seditious papers which feed the fire on which revolutionary conspirators cook up a 'Charm for a Democracy' in Rowlandson's eponymous 1799 print. On the one hand, such habitual representation of Darwin in reduced and caricatured form suggests that such detailed engagement with the content of his ideas as even *The Golden Age* represents could no longer be countenanced, but on the other, the focusing of hostility to a vitalist science of nature onto the personal figure of Darwin himself arguably enabled vitalist ideas to re-emerge in relatively uncontroversial form at a later date, as they were to do with Abernethy's interpretation of Hunter in the second decade of the next century. Darwin himself, who regarded the most controversial chapter of his *Zoonomia*, on evolution, as a 'barrel' with which to amuse the whales (in other words, as a lightning rod for the criticism which might otherwise damage the rest of the work), would perhaps have regarded such a pay-off – the sacrifice of his reputation for the survival of his ideas – with some equanimity, and perhaps even with an amused acknowledgement of how a theory of nature which emphasised the powers of spontaneous regeneration could itself manifest such phoenix-like qualities.

In light of the way the figure of Darwin himself stands in metonymically for his ideas in later attacks, it is interesting to note *The Loves of the Triangles'* dependence on a caricatured version of the author in its earlier satirical sallies. The introductory letter, supposedly from the poem's author, a Mr Higgins, proffering the work for publication in the journal, performs some important spadework in establishing the ideological targets for the poetic parody which follows; another letter, following the final extract from the poem, reiterates the same points.[44] Identifying a progressivist politics beneath the natural philosophical surface of *The Botanic Garden*, Canning makes use of Higgins' letters to expose the implications of an account of a self-determining nature which evolves and progresses through its own energies and actions if applied to the sphere of human activity. The essential character of Darwin's nature – thrusting, self-seeking, self-determining and

Figure 3 James Gillray, *New Morality; – or – the promis'd instalment of the high-priest of the Theophilanthropes, with the homage of Leviathan and his suite*, Anti-Jacobin Review and Magazine; or Monthly Political and Literary Censor, no. 1 (1 August 1798), 1

Figure 4   James Gillray, *The Apples and the Horse-Turds; – or – Buonaparte among the golden pippins*

self-regulating – had an evident radical significance if applied to human society; Canning's attribution to Higgins of an earlier poem, *The Progress of Man*, is both a jibe at Richard Payne Knight's *The Progress of Civil Society* (1796), and a sharp anticipation of the ways Darwin's vision of nature could be extended into an account of human social development itself – as indeed it explicitly would be in *The Temple of Nature*. Following the announcement of his 'first principle' – the inverted Popean maxim, '*Whatever is, is* WRONG' – Higgins outlines his second belief in the '*eternal and absolute* PERFECTIBILITY of MAN', arguing that 'we have risen from a level with the *Cabbages of the field* to our present comparatively intelligent and dignified state of existence, by the mere exertion of our own *energies*'. Were it not for the repressive effects of 'KING-CRAFT and PRIEST-CRAFT, and the other evils incident to what is called Civilized Society', continued exertions would 'raise Man from his present biped state, to a rank more worthy of his endowments and aspirations; to a rank in which he would be, as it were, *all* MIND, would enjoy unclouded perspicacity and perpetual vitality; *feed* on OXYGENE, and never DIE, but *by his own consent*'.[45] Higgins' (or Darwin's) belief in the 'free agency of man's *natural intellect* and *moral sensibility*' is thus presented in travestied form as a ludicrous and unachievable utopianism, sustained only by the unfounded scientific speculations of a provincial eccentric and fantasist. The danger of mapping a philosophy of nature onto a philosophy of man is underlined too in Higgins' second letter, where a declared interest in natural regeneration becomes a desire to undo all existing social and governmental structures, an expression of revolutionary principles made explicit in the characterisation of the French Revolution as the 'great experiment of regeneration'.[46] Whereas, when the poem itself gets underway, the looseness of the connections between the various themes, topics and scenes it relates works in a general way to manifest Higgins' dunce-like madness, here the same readiness to bridge differences has the specific effect of demonstrating the danger of leaping too readily from a science of nature to a science of politics. Just as the reviews would later disparage Darwin's attempted hybridisation of poetry and science, so too does Canning implicitly reject a politics founded on natural philosophy – especially a natural philosophy which he sees as simply a 'sort of cover and disguise' for 'securing the favourable reception' of Darwin's sentiments.[47]

With its disinclination to favour Darwin's science with too detailed an investigation, Higgins' faith in a principle of 'perpetual vitality' attained through oxygen is the nearest *The Loves of the Triangles* comes to an explicit representation of vitalist nature. Since Priestley's claims, in *Experiments and Observations on Different Kinds of Air* (1775), that the manufacture of improved aerial environments, thought a possible result of his pneumatic experiments, would have beneficial moral and social effects, 'dephlogisticated' air had been associated with progressive politics; Canning's reference to oxygen thus invokes not merely a contemporary scientific discovery but the politicised claims made and applications sought for it. Darwin himself favoured Lavoisier's term 'oxygen', as well as his alternative chemical methods and definitions over those of Priestley, and was one of the first to use the word 'oxygen' in print – including in *The Economy of Vegetation*, where it is given its first positive usage.[48] He was an enthusiastic supporter of Beddoes' investigations of the therapeutic uses of oxygen to cure disease by providing 'vital energy', and discussed developments in such pneumatic treatments in the second volume of *Zoonomia* (1796). Darwin's declaration of Lavoisier's oxygen as 'revolutionary' was largely a response to the breakthrough represented by his alternative chemical methods and terms, but Darwin's association with oxygen in the public mind, especially given the recognition of the roles played by Lavoisier and other chemists in revolutionary France, gave the term a more immediately political charge.[49] Higgins' enthusiasm for 'feeding on oxygen' thus links Darwin directly to the revolutionary applications of contemporary science, as well as reiterating the spurious and fantastical nature of the claims made by scientists for what their discoveries might achieve. Through this association with oxygen, which was also commonly called 'vital air', theories of vitality thus suffer a similar debasement: produced through the unfounded claims of experimental philosophy, and seen as manifestations of wild, potentially dangerous, political sentiments. In this they share in the characterisation given to many forms of natural philosophical or 'scientific' learning by loyalist propaganda, and especially in Burke's influential *Reflections on the Revolution in France*; Canning's choice of geometry as the particular focus of Higgins' poetic system, for instance, echoes Burke's attacks on the geometrical principles followed by the French revolutionaries. Whilst Darwin's vitalist theories are attacked in *Loves of the Triangles*, then, in part they are recipients of the general hostility directed at the inhumane, instrumentalist thinking of, in Burke's terms, all 'sophisters, economists and calculators'.

In fact, as with other satirical responses to Darwin, *Loves of the Triangles* is happier attacking Darwin (and by implication his science) on quite another front: his flouting of the proper modes of literary production and performance. This not only carried consequences for the reception of Darwin's final work, *The Temple of Nature*, reviews of which echoed the criticisms implicit in Canning's attack, but also contributed to the ongoing construction of Tory versions of literary culture – a project again continued in the pages of the periodical press through into the next century. The *Dunciad*-model followed by Canning (it had echoed too in *The Golden Age*) enables him to associate Higgins with Grub Street excesses of senseless literary output, manifested not just in the unorganised prolixity of *The Loves of the Triangles* itself, and Higgins' proposed subsequent project of a drama along the lines of German *Sturm und Drang*, but in the endless flow of '*Encyclopedias, Treatises, Novels, Magazines, Reviews,* and *New Annual Registers*' which it is seen to be part of.[50] Such literary unrestraint is presented as the textual manifestation of Higgins' restless but redundant intellectualising, and his uncontrolled mania for system-making; it receives its visual equivalent in the crowded and chaotic scenes of Gillray and Rowlandson's graphic satires, where the radical rabble makes a disorderly progress to worship at the shrine of revolutionary deities (*New Morality*) or is heaped together in a confused mass in a steaming and indistinguishable pile of republican excrement (*The Apples and the Horse-Turds*). The encyclopaedic inclusiveness of Higgins' work, which aligns the philosophically recondite (algebraic maxims and deities) with the banally prosaic (a celebration of the spit-roast), or, in its extensive annotation, alternately combines empty speculation, questionable assertions, bizarre theory or minute and self-evident truths, suggests a fundamental failure of any system of literary or intellectual order or hierarchy. Higgins' desire to find equivalences or connections between, say, 'Mr. Gingham and his daughters overturned in a one horse chaise' and 'the Three Curves, Parabola, Hyperbola, and Ellipsis', not only produces a comically disordered poetic narrative, but suggests the literary and intellectual failure of the democratic instinct in which it must be assumed to originate.

Faced with such chaotic scenes, Tory satirists place a new premium on order, value, judgement and discrimination, and identify the space of writing itself as the place where such principles and hierarchies are to be determined and enacted. Thomas Mathias' *The Pursuits of Literature* (1794–97), with its extensive survey and evaluation of the contemporary literary scene, is one manifestation of such urges; it was praised as a model of satirical practice in Canning's *New Morality*, and proffers a swipe at Darwin's feminine verse – his 'flimsy, gawzy, gossamery lines' – as well as reiterating the charge of his poetry's sexual suggestiveness.[51] The poem's sense (expressed in Popean cadence) that Priestley, for instance, 'writes on *all* things, but on nothing well', suggests that in the current climate of heated political divisions a new insistence on specifically literary evaluation and judgement will have an ameliorative effect on the damaged national psyche: that a reinvigoration of aesthetic practice will both counter dangerous political sentiments and restore proper balance and order to social as well as literary actions. The poem thus at once attempts to demarcate a sphere of aesthetic activity and literary judgement separate from a political one, and, contrarily, to insist on the ameliorative powers of the former on the latter. Canning's *New Morality*, which similarly opens with a survey of the current literary scene, this time focusing on the state of satire itself, is at least overt about the political purpose it envisages for the newly invigorated satire which it seeks. In a more philosophically oriented poem than the parodic *Loves of the Triangles*, Canning turns to a heroic and patriotic version of satire to defend the nation in a moment of national crisis. In a text which is as much about poetry itself as about the 'New Philosophy of Modern Times', as becomes evident in its closing pages, a ruggedly masculine, public and principled poetry is hailed as the successor to Burkean patriot speech – in implicit contrast with the supposedly light, feminine, domestic and privately oriented entertainments of Darwinian verse.[52]

Interpretative responses to *New Morality*, perhaps informed consciously or otherwise by the Gillray engraving inspired by the poem, often emphasise the text's concern with political philosophy, and its tripartite attack on philanthropy, sensibility and justice – qualities later depicted as three revolutionary deities in Gillray's print.[53] The poem's treatment of such topics, however, cannot be fully separated from its sense of its own purpose as an attempt to reinvent the morally engaged Popean satire of 'a milder age' for the new challenges of the current one. Whilst Darwin, who was to make an appearance in Gillray's illustration (with a basket of revolutionary flowers on his head, labelled 'Zoonomia, or Jacobin Plants'), is not named in Canning's poem, his ideas appear at a critical moment in the text's discussion of sensibility, and provide the occasion not simply for an attack on the dubious and corrosive sensibility Canning sees as championed by Darwinian accounts of vegetable sensitivity, but also to differentiate his own poetic practice from that of Darwin (see Figure 3 and jacket). The personified 'Virtue' of Sensibility makes her appearance in the poem through lines which directly invoke Darwinian accounts of vitally animated nature, as well as mimicking his poetic practice:

> Sweet SENSIBILITY, who dwells enshrin'd
> In the fine foldings of the feeling mind; –
> With delicate *Mimosa's* sense endu'd,
> Who, shrinks instinctive from a hand too rude;
> Or, like the *Anagallis*, prescient flow'r,
> Shuts her soft petals at th' approaching show'r.[54]

With their use of personification, a narrative focused on plant life and attribution of human qualities to vegetables the lines could almost be mistaken as by Darwin himself, but the assumption of Darwinian style and the apparent praise of sensibility are equally short-lived, as the former is replaced by an admonitive mini seduction narrative, and the latter is shown to be politically corrupted and morally degraded. In the lines which follow, sensibility is shown to have been seduced and miseducated by Rousseau, so that sensibility's judgements are 'False by degrees, and exquisitely wrong'. In feeling more for the 'crush'd Beetle' and the 'widow'd Dove' than for parents, friends, king and country, sensibility represents the dangers of excessive femininity, which is to be countered not only in the poem's celebration of 'fix'd' principles, but in the heroic manliness of its own style: the 'manly vigour' and 'patriot warmth' which, in contrast to the effete flowery Darwinianisms, stamp its 'nervous' lines. Such qualities eventually find their apotheosis in the celebration of the recently deceased Burke's combination of Greek and Roman virtues. Only by echoing Burkean oratory, with 'antient Morals, antient Manners' and 'manlier virtues' constantly in view, can the Muse's 'thunder cloath'd' verses secure the nation a safe passage through the current 'storm'.[55] If, equally, the revolutionary threats of the 1790s demand a rediscovery of Pope's poetic legacy, it is a testosterone-heavy reinvention of high satire, rather than the lighter Popean fancies invoked by Darwin as his poetic model, which are in order. If the lines on Sensibility represent a brief assumption of the latter, it is an act of poetic cross-dressing only indulged to maximise the effects when such feminine coverings are dramatically discarded.

As with the reviewers' later invocation of disciplinary or discursive boundaries unallowably transgressed by Darwinian verse, *New Morality* and *The Pursuits of Literature* again show that it is through renewed attention to, or reinvention of, the proper forms of writing that the loyalist response counters the threats posed by radical natural philosophy and progressivist politics. Whilst on the one hand this produces a robust articulation of a Tory vision of poetry's purpose and identity, including its place in the cultural life of the national polity, on the other, as has already been suggested, its focus on poetic practice leaves the door open for the return of vitalist science, or other philosophical controversies, by other means. Whilst their strategy of focusing on literary decorum produces a reinforced articulation of the importance of literary judgement, order and value, the question remains as to the effectiveness of this invocation of disciplinary regulations as a means of containing or countering the threat of scientific ideas which are only occasionally expressed through literary discourse. The retrospection evident in the models to which Canning, Mathias and others turn – primarily the satire of Pope, and especially *The Dunciad*'s attack on Grub Street's disordered and excessive literary culture – may hint too at their inadequacy to fully combat the particular challenges of the current age. Like Mathias' *Pursuits of Literature*, *The Dunciad* primarily aims to expose the literary, moral and mental failings of those who would attempt to compete with Pope in specifically literary efforts, but the challenge of countering the import of Darwin's writings lies only partially in addressing the value or otherwise of his poetic performance. As we have already seen, understood most fully, the Darwinian project of 'enlisting the imagination under the banner of science' implies the dissolution of the strict disciplinary boundary between literature and science which the Tory satirists would reinforce, through a recognition of the mental activities – primarily the free-flowing mental speculation of reverie – common to both. Faced with the challenge of such a vision, the harking back of Canning and others to the satire of 'a milder age' appears retroactive, especially when compared to the ways Coleridge – no fan of Darwin's poetry, as we have seen – was influenced by his accounts of imaginative reverie in the development of his own theory of the poetic imagination.

Such a 'high' Romantic assimilation of Darwin represents one route in which the poet continued to exert a cultural influence despite the repressive efforts of loyalist satirists in the 1790s.[56] Despite their attempts to rethink the identity, role and place of literary culture in a new era – as for instance in Wordsworth's 'Preface' to the *Lyrical Ballads*, which is already responding to the incipient challenges of both modern science and emergent democratic sentiments – such efforts arguably resulted not in the dissolution of such discursive boundaries as Canning and others invoked, but in the formulation of new norms of literary value, aesthetic practice and canonicity. The cultural status ascribed to both Wordsworth and Coleridge in their later years (the former becoming poet laureate, for instance), as well as the adoption of Romantic aesthetics, canons and principles in the numerous institutionalisations of 'Literature' during the nineteenth and twentieth centuries are two signs of this. But Darwin's inclusion in an alternative, radical, literary practice in the 1790s represents another form of his cultural assimilation, even if it was to be short-lived. Whilst Canning and others were attacking Darwin in the high- and middle-brow periodical press, extracts from his writings were adopted and reprinted in quite different cultural arenas: in the 'creative anthologies' produced by agitators for political reform, including Daniel Isaac Eaton's *Politics for the People*, and in the political speeches of John Thelwall.[57] If such publications and their sentiments were eventually suppressed by repressive government legislation (despite the acquittal of Thelwall and others at the treason trials of 1794), they nevertheless represent an alternative mode of literary and textual practice which exploited the figurative possibilities of a vitalist science of nature in productive and suggestive ways. In doing so, they sketched an alternative vision of the relationship between science, politics and culture in which, in place of enforced divisions between distinct areas of activity, a 'creative' disciplinary looseness was a central weapon in the repeated elaboration of arguments for reform.

Darwin and the radicals

Amongst the profusion of radical journalism produced in the mid-1790s, Daniel Isaac Eaton's *Politics for the People* is notable for its fervour and for its comparative longevity. Initially named *Hog's Wash, or, A Salmagundy for Swine*, it was a weekly publication designed to foster support among the lower classes for the causes of parliamentary reform and universal suffrage; it ran for 60 issues from September 1793 to March 1795 – surviving Eaton's prosecution for seditious libel in late 1794. Like Thomas Spence's *Pig's Meat*, the paper was largely composed of short extracts from a wide range of literary, political and historical sources, mixed with original material including fables, songs and radical verses. Swift, Harrington, Dr Price and Godwin thus found space alongside Goldsmith, Johnson, Churchill, Dryden and Rochester; Milton and Shakespeare could also occasionally be found. Extracts chosen were those which could be read to reinforce radical or liberal political sentiments, or lend themselves to easy application to the current political moment – a process aided at times by the addition of interpretative pointers, such as italics or titles. A reprinting of Gray's *Elegy in a Country Churchyard* in *Pig's Meat* thus underscores the text's class politics through emphasising such words as 'tyranny', or an extract from Goldsmith's *The Deserted Village* is titled 'A Lamentation for the Oppressed'.[58] From one perspective, the cheek-by-jowl nestling of such a heterogeneous collection of writings may have appeared to represent precisely the unorganised and chaotic mix of literary, political and intellectual culture feared in the satirists' invocations of Grub Street, or represented in graphic satires – where the very figures who share the pages of Spence or Eaton's radical papers can be seen occupying the same visual space. From another perspective, however, the cumulative effect of such a collected mass of evidence is to demonstrate the wealth of literary, critical and political material from which criticisms against the established government, anti-monarchical sentiment and arguments for political reform might be drawn. At the same time, the claiming of such sources for a largely lower-class readership constitutes the kind of assertion of participation in a 'republic of letters' as commentators such as Paul Keen have seen as central to working-class radical politics at this time.[59] It also constitutes an entirely different attitude to the existing literary heritage from the Tory satirists' impulse to protect and preserve the established identities and boundaries of literary practices. Whilst their attacks on Darwin illustrate a resistance to new forms of literary production, the overriding drive for inclusiveness in such publications as *Politics for the People* or *Pig's Meat* effectively rewrites established literary and political canons by foregrounding those authors and works whose writings can be exploited for the compiler's particular political purposes. Against a defensive Tory model of the place of the literary in national political life – where energy is primarily expended on the 'policing' acts of discriminating, excluding and rejecting – such radical publications manifest an alternative, democratic, as well as obviously politicised attitude to the existing body of political, historical and literary writing. The editorial principle apparently suggested by the sheer variety of sources collected by the compiler – that all forms of writing might potentially be included – is only modified by the obvious preference for that which contributes to the radical political agenda. In a radical textual practice which implicitly challenges painstaking attempts to maintain literary hierarchy and generic difference, value is here determined not by reference to established literary principles or textual decorum, but according to exegetical ease and political utility.

In Marcus Wood's description of such publications as *Politics for the People* as 'creative anthologies', creativity is best ascribed to the act of collecting, rather than reading, the material which has been gathered together: the heterogeneity of the sources from which Eaton's content is culled is not to be matched with a parallel diversity in the act of interpreting it. With an implicit premium placed on the particular interpretative use to which each extract is to be put, there is little room for the excessively metaphorical or figurative; as Michael Scrivener has demonstrated in his study of Jacobin writing, allegory, with its relatively close connection between vehicle and tenor, is a particularly favoured literary form, closely followed by the often similarly exegetically straightforward fable.[60] This means that, where extracts from Darwin are used, they are carefully chosen for their obvious political applicability. This is evidently the case when *Politics for the People* reprints a passage from *The Economy of Vegetation* to open its twelfth issue. Taken from the second canto, which addresses topics associated with earth (the other three cantos address the elements of fire, water and air), the lines follow from a discussion of coal, amber and Franklin's electrical experiments to a celebration of the spread of liberty, metaphorically presented as electrical contagion, in America and Ireland. The awakening of the 'giant' of France through the touch of the 'patriot-flame' – an act of electrical resuscitation which was to be disparaged in a footnote to *The Loves of the Triangles* – brings the passage to a climax, and provides an image of political invigoration envisaged as a renewed natural strength against which the petty chains of repression are powerless:

> – Touch'd by the patriot flame, he rent amazed
> The flimsy bonds, and round and round him gazed,
> Starts up from earth, above the admiring throng,
> Lifts his Colossal form and towers along;
> ...
> Calls to the good and brave, with voice that rolls
> Like heaven's own thunder round the echoing poles;
> Gives to the winds his banner broad unfurl'd,
> And gathers in its shades the living world.[61]

Carefully extracted and relocated in the pages of Eaton's journal, Darwin's metaphorical connections between the natural phenomena such as electricity which are the immediate subjects of his poem, and contemporary political events, do not appear so different from the allegorical tendencies of much Jacobin writing. In turn, his presentation of the rise of political liberty as an inevitable function of natural power draws on the authority of contemporary natural philosophy to offer an appealing prophecy of a newly liberated 'living world'.

In this extract, the figurative links between the natural and political worlds are made by Darwin himself, but despite his private sympathies for progressive political causes, such moments are relatively rare in his verse. When Thelwall quotes from a later section of *The Economy of Vegetation* to end his *Sober Reflections*, the connections between nature and politics have to be drawn by Thelwall himself. Turning this time to the poem's final canto on air, Thelwall's attention is caught by lines which, following a description of the ways stars, suns and other astronomical systems 'too must yield to age' and become 'extinct', celebrate the re-emergence of 'Immortal Nature' in renewed, if changed form:

> – Till o'er the wreck, emerging from the storm,
> Immortal Nature lifts her changeful form,
> Mounts from her funeral pyre on wings of flame,
> And soars and shines, another and the same.[62]

But the lines as they appear when Thelwall reprints them are subtly different:

> High o'er the wreck, emerging from the storm,
> Immortal FREEDOM lift her changeful form;
> Mount from the funeral pyre on wings of flame,
> And soar and shine, another and the same![63]

Darwin's iteration of a favourite theme, the repeated regeneration of nature, in an indestructible cycle of constant return, becomes silently adapted in Thelwall to describe an inevitable re-emergence of liberty – a theme fitting to Thelwall's insistence, as *Sober Reflections* reaches its climax, of the exertions of reason leading to justice and national renewal, but quite distinct from Darwin's more abstract concerns. Darwin's precise natural philosophical vision of a regenerative nature becomes, in Thelwall's adaptation, the celebration not of the energies of abstract 'nature' but of the specific exertions of human effort, reason and will. It is through deliberately applied human force, not vaguely invoked natural power, that the utopia of political freedom will be achieved: liberty will be achieved as a result of a conscious and collective intervention in material reality, not as an automatic and inevitable act of fate. In this way, Darwin's predilection for invoking the powers of abstract natural energies is refigured to articulate the effects of combined human efforts; at the same time, the ongoing historical human struggle for justice is presented as an inevitable force of nature, a manifestation of the will or fate of the universe. In a messianic narrative, reason is glossed as a vital energy of nature, and justice as experiencing an inevitable return.

In some ways, Thelwall's readiness to silently adapt Darwin's text to his own ends represents the same 'creative' attitude to literary culture as Wood identifies in the compilations of Eaton and Spence. It manifests a willingness to find political expediency in whatever sources can be culled from the contemporary moment – and in a strange way explains, if it does not justify, the tendency of government agents to regard Thelwall's possession of a copy of Darwin's *Botanic Garden* as seditious (Thelwall complains elsewhere of how this volume, together with Godwin's *Political Justice*, had been taken during a raid on his home).[64] That act of the physical removal of supposedly suspicious texts represents the logical extension of the Tory satirists' impulse to police and repress politically suggestive forms of writing, but its absolutism fails to address the adaptive resourcefulness of Thelwall and others, who will find grist to their mill in whatever literary or other sources they come across.[65] Their continued interpretative capacities suggest the impossibility of attempts to separate a literary culture from a political one: their very ability to find figurative meaning in the most abstract of sources would present the most unlikely text as a fund of potential political meaning. But in this meeting of political rhetorics and textual sources, interpretative creativity works at a number of levels: not only to extract immediate political meaning from literary (or natural philosophical) words for current political purposes, but also to project such interpretative possibilities back into the text as a potentially more permanent meaning. Darwin's vitalist natural philosophy thus retained its aura of controversy well beyond the era of Eaton and Thelwall, tinged as it was both with their highlighting of the radical applicability of his faith in natural inevitable energies, and with the residual suspicions of Canning's and other attacks. The reputation of the poet of regeneration and vital energy would thus prove less capable of resurgence and reanimation than the nature which he celebrated – but the vitalist science which he expounded nevertheless lived on to become a significant force well into the early decades of the next century.

6
Animation and Vitality in Women's Writing of the 1790s

Creating origins

> Did I request thee, Maker, from my clay
> To mould Me man? Did I solicit thee
> From darkness to promote me?
> > *Paradise Lost* X 743–5, epigraph to
> > Mary Shelley's *Frankenstein* (1818)

'Hateful day when I received life! ... Cursed creator! Why did you form a monster so hideous that even you turned from me in disgust?'
> > Mary Shelley, *Frankenstein*[1]

> The fairest of created works was made
> To share, with man, th' empire of creation,
> T'enjoy its comforts and its sweets, its pains
> And suff'rings
> > Anon., epigraph to Mary Ann Radcliffe,
> > *The Female Advocate* (1799)

> Wherefore are we
> Born with high Souls, but to assert ourselves?
> > Epigraph to Mary Robinson,
> > *Letter to the Women of England* (1799), from
> > Nicholas Rowe, *The Fair Penitent* (1703)

'I know I am in some degree under the influence of a delusion – but does not this strong delusion prove that I myself "am of subtiler essence than the trodden clod:" ... have I desires implanted in me only to make me miserable?'
> > Mary Wollstonecraft, *Mary* (1788)[2]

For what am I reserved? Why was I not born a man, or why was I born at all?

Mary Wollstonecraft, *The Wrongs of Woman* (1798)[3]

Every thing must have a beginning

Mary Shelley, 'Introduction', Standard Novels Edition of *Frankenstein* (1831)[4]

'Every thing must have a beginning': Mary Shelley's words are offered as, a decade and a half after *Frankenstein*'s publication, the author of one of the most famous novels about beginnings and their consequences attempts to tell the story of *Frankenstein*'s own origins. As the first two epigraphs above suggest, *Frankenstein* is a story in which acts of creation themselves are questioned, even rejected, but placed as they are above, alongside quotations from women's writings of the 1790s, including by her mother, Shelley's questions about creation begin to look like they had already started some time before. The questioning of his maker by Frankenstein's creature echoes the turn to myths of creation by women writers of the 1790s, as they asserted the rights of women against political and social injustices rendered newly challengeable in the climate of debate ushered in by the French Revolution. For Mary Ann Radcliffe, Mary Robinson and Mary Wollstonecraft, one way such challenges might begin is with creation. The nature of creation, the history of women's place in society and the origins of their social oppressions: such concerns appeared to hold the key to some of their questions about women's current state, and the possibility of ameliorating it. In part, such lines of enquiry manifested an analytic reason characteristic of radical thought of the time, which repeatedly sought to trace effect to cause, to lift the curtain of social appearances and reveal 'things as they are' – a method of thought condemned in Edmund Burke's denouncement of a revolutionary determination to tear society into a 'chaos of elementary principles', and at odds with his own desire to uphold such social fabrics as already existed.[5] But it is also noticeable how often these women's narratives of origins can themselves only get started by borrowing the words of others – of the men who have gone before them. Shelley's epigraph reminds her readers of the importance of Milton's *Paradise Lost* as a foundational text for her own story; Robinson's epigraph from Nicholas Rowe's domestic tragedy, *The Fair Penitent*, and Wollstonecraft's quotation from Edward Young's *Night Thoughts* play the same trick of framing their own questions through quoting and reworking already-existing male speech. Every thing must have a beginning, as Shelley says (and, as she acknowledges, the phrase already belongs to Cervantes' Sancho Panza), but in some cases, it appears, the men have ensured that that beginning has already begun.

The men have already got the story going too in Shelley's narrative of the origins of *Frankenstein* itself. In her now-famous account of the Switzerland holiday of 1816, when a 'wet, ungenial summer' led to Byron's call for ghost stories from each member of the party, Shelley is only rescued from her lack of inspiration by a night-time reverie triggered by listening to conversations between Byron and Percy Shelley on the 'doctrines' of the 'principle of life'.[6] Identified as the true beginning of her fictional creation, her narrative thus takes a science of animation, infused through the dreams and imagination of the novelist, as its own generative principle. Taking its cue from Shelley's account, much critical work of recent years has further elaborated the relationship of Shelley's novel to contemporary sciences of life. Such work has highlighted, for instance, the contemporaneous 'vitalist' debate between John Abernethy and William Lawrence; the biographical links between Lawrence and Percy Bysshe Shelley; the presence amongst the holiday party of Byron's physician, Edinburgh-educated John Polidori; and the discussions, by those present at the Villa Diodati, of vorticellae experiments reported by Erasmus Darwin, which appeared to show the spontaneous generation of life.[7] But for all the precision with which *Frankenstein*'s original relationship to such a context might be plotted, the science of animation is usually left far behind in the voluminous interpretations the novel itself has attracted, in which larger, looser questions – such as the nature of responsibility, knowledge, sociability, friendship, and modern science in a more general sense – might figure alongside discussion of the text as a response to the French Revolution, or to Shelley's fraught experiences of motherhood. If anything, the specificity of *Frankenstein*'s origins in the precise confines of a particular contemporary scientific question (as some critics, at least, would have it), is, strikingly, inversely proportional to the extraordinary, almost unparalleled range of interpretations the text has attracted: the precise narrowness of the moment and constituents of its inception balanced by the breadth of what it has appeared to generate in its exegetical afterlife. In this, of course, it echoes the move from specific origin to unforeseen consequence of Victor Frankenstein's own creation – or perhaps, indeed, of any life form.

But for all the attention paid to the historical moment of *Frankenstein*'s inception, it is worth noting that the scientific debate which, in this critical story at least, triggers *Frankenstein*'s birth, although given a new public profile by the 1814–19 Lawrence/Abernethy vitalism controversy, essentially revolves the same questions as had been circulating since at least the mid-eighteenth century: the origin and nature of the life-principle, and whether life is a quality of matter itself or something superadded to it. In the public confrontation between Abernethy and Lawrence, couched in terms of a debate over the proper interpretation of John Hunter's work, different answers to these questions were embodied in the polarised figures of Abernethy, with his reassuring conciliation to current religious anxieties, and Lawrence, with his controversial materialism – from which he was eventually forced to retreat; the significance of the debate itself was understood to be more than the settling of an abstruse scientific question, but to carry far-reaching implications for contemporary social, political and religious orthodoxies.[8] But Abernethy's position as stated at his Hunterian lectures at the Royal College of Surgeons in 1814, that life must be 'superadded' to matter, varies little from the conclusions reached in Thelwall's 1793 *Essay Towards a Definition of Animal Vitality*. The conversations between Byron and Shelley on the 'principle of life' overheard by Mary Shelley in 1816 could have taken place ten or 15 years earlier; after all, Darwin's reports of the vorticellae experiments, published in his *Temple of Nature* in 1803, simply extended his interest in animal and plant vitality first made evident to the reading public in his *Botanic Garden* of 1789.[9] In this sense, the scientific context for *Frankenstein* reaches back far beyond the moment of its initial inception to the 1780s and 1790s, and given this, it is perhaps significant that Shelley backdates the action of her novel to the close of the preceding century.

As this chapter will show, the period of the last years of the eighteenth century, in which Shelley sets the action of her novel, saw similar interest in creation and animation circulate in and inform literary and political writing by women, as inform Shelley's later text. But in the 1790s, a concern with animation and vitality, which is evidently at some level informed by contemporary biological and physiological science, is less clearly demarcated as 'scientific' than in *Frankenstein*. Where, in that text, the science of generation swiftly and straightforwardly generates the larger story, in women's writing of the 1790s the relationship between a scientific language of animation on the one hand, and 'literary' questions and concerns on the other, is at once less overt, more submerged and more interconnected, as discourses of animation, vitality and life (as well as their opposites, stagnation, stupefaction and death) are repeatedly deployed to address these writers' preoccupations with the contemporary social construction and oppression of women. As the epigraphs to this chapter have already suggested, such discourses are present when writers, whether fictional or non-fictional, trace their concerns to first causes, to question the nature of God's creation, to ask after the meaning of their life or animation, and to solve the riddle posed by women's possession of qualities (rational thought, souls, desires, feelings) seemingly little acknowledged by current social custom. In Mary Robinson's *Letter to the Women of England*, the assertion that the 'sex of vital animation' is as yet unknown bolsters her rejection of an established and oppressive system of gender differences.[10] Robinson's phraseology, inflecting contemporary debates about the cause, nature and identity of vitality, offers a more physiologically attuned version of the eighteenth-century philosophical and theological commonplace, that the 'soul' or the 'mind' has 'no sex'; in doing so, a bland formulation of received wisdom is transformed into sharper, more confrontational, more current form. And within fictional writing, the opposed tropes of deathliness and vitality, stagnation or suffocation and life, torpor and energy are repeatedly deployed both to suggest the oppressions endured by a generation of literary heroines, and to begin to sketch, through a language of animation and energy, how women's own vital resources may provide some kind of solution. Such animated energies take different forms in different texts – from the vitality associated with Sibella in Eliza Fenwick's *Secresy* (1795), to the genius and sensibility emphasised in Mary Wollstonecraft's two fictions – and also, as the rest of the chapter demonstrates, produce a variety of outcomes, successful and otherwise.

A relatively straightforward example of how animation, in one of its guises, is invoked as the panacea for the accumulated troubles of one woman's life, is given in the story of Fidelia, a short sentimental fiction which concludes Mary Ann Radcliffe's *The Female Advocate: or an Attempt to Recover the Rights of Women from Male Usurpation* (1799). Most of Radcliffe's text rehearses a familiar litany of the difficulties faced especially by middle-class women, educated to a station which, following the death of a husband or father, too often they can no longer themselves independently sustain. In a curious mixture of pragmatism and protectionism, Radcliffe both calls for tradition-ally female jobs encroached on by men, such as shop work and hairdressing, to be returned to women, and argues for more charitable intervention on behalf of distressed women, from those best placed to perform such benevo-lent acts. If such measures fail, women have little option, Radcliffe suggests, but to sell themselves for their own survival, a fate illustrated in a turn from prose to sentimental verse which transforms her practical exposition of day-to-day difficulties into the high drama of moral and existential crisis:

> In that dread moment, how the frantic soul
> Raves round the wails of her clay tenement,
> Runs to each avenue and shrieks for help,
> But shrieks in vain. How wishfully she looks
> On all she's leaving, now no longer hers,
> A little longer, yet a little longer.[11]

The turn to prostitution, that familiar tragedy of sentimental and didactic narratives, makes evident the precarious existence, caught between life and death, of the women whose difficult circumstances Radcliffe narrates, and emphasises that precarious state as a faltering on the brink of not only literal but moral death too.

The imminent tragedy faced by the nameless fallen woman here is paral-leled in the frontispiece to Radcliffe's text, an engraving which shows the protagonist of her later story, Fidelia, hovering on the bank of a river, on the point of throwing herself in to her death (Figure 5). As well as speaking, in its prominent position at the head of her text, for what Radcliffe evidently understands to be the perilous situation of many women, poised on a nar-row path between life and death, the picture also illustrates the climactic moment of Fidelia's own narrative. Collecting together a number of the themes and dangers on which Radcliffe has touched in the earlier discussion, Fidelia's tale tells the story of the daughter of a free-thinking gentleman,

Figure 5 Frontispiece to Mary Ann Radcliffe, *The Female Advocate*, 1799

educated in morals but not religion. On his death, having rejected marriage to a man she does not love, she succumbs to the seductive advances of Sir George Freelove, only to be abandoned after a year, penniless, homeless and friendless. Meditating her dismal situation, she resolves on suicide, and is only saved from her plunge into the river by the quick action of an onlooker, an aged, impoverished gentleman, who takes her into his home. There she meets his wife who, though living in poverty, dying of cancer and having buried eight of her children, nevertheless has eyes 'animated with joy', a description which introduces the central theme of the narrative's resolution. Reiterating her earlier solitary musings on why we are 'animated to suffering', Fidelia asks the wife 'how it is possible for you to preserve, amidst such complicated misery, that appearance of cheerfulness and serene complacency which shines so remarkably in your countenance, and animates every look and motion'. The woman's response is that her religion is her strength, and her subsequent teaching of Fidelia in the lessons of the Bible leads to Fidelia's eventual decision to support herself by going into service, sustained in her humble life by the comforts of her faith. 'Virtue itself', the narrative concludes, 'cannot confer happiness in this world, except it is animated with the hopes of eternal bliss in the world to come.'[12]

Fidelia's story reiterates many of the problems faced by women which writers like Radcliffe addressed – the fate of educated daughters lacking independent financial support; the dangers faced on the death of fathers; the sexual double standard whereby libertine males pursue a sexual freedom which condemns women as social outcasts; the difficulties involved in rejecting a marriage partner chosen by a guardian. But despite the practical measures suggested in Radcliffe's earlier discussion, the story's only solution to these multiple difficulties is a life 'animated' by religious faith – the word 'animation' or its cognates is repeated at least three times in the final few pages. The term 'animation' operates very differently here from the examples which will be explored later in this chapter – signalling, for all its suggestion of liveliness, a stoic submission to the sufferings of life, or, more precisely, a vitality paradoxically sustained via such stoic humility – but it is nevertheless striking that it is only by returning to a fundamental insistence of the power of vitality (albeit one provided by religious faith) that Fidelia's concerns are resolved. As we will see, alternative versions of animation and vitality, deployed to develop ideas of genius, sensibility and creativity, and to explore an interest in human nature which can encompass concerns as various as contemporary physiology on the one hand, and the relation between body and soul on the other, operate in a different vein in the more extended literary fictions of Eliza Fenwick and Mary Wollstonecraft. But for all these writers, it would seem, whatever their differences, some form of animation, vitality or energy – desired, threatened, defended, demanded – remains a central touchstone of their work.

Of course, preoccupation with animation and transport is a recognised feature of much Romantic writing, one which is usually assimilated to

Romantic aesthetic concern with imagination and creativity. So dominant is this critical theme, in fact, that it threatens to overshadow, or co-opt, the concern of earlier eighteenth-century natural philosophy with the forces which animate nature and matter. This represents an alternative context within which to consider the works of early Romantic-period writers Eliza Fenwick and Mary Wollstonecraft – although, as I shall go on to suggest, such contexts have their limitations as well as their uses. Unlike John Thelwall or Erasmus Darwin, neither writer can be shown to be explicitly and deeply involved in controversies over vitality, but there is much at a general level to point to their likely knowledge of such a significant debate in contemporary physiology. Both were working at a time when, first, the language of vitality and animation filled the periodical press, and, second, natural philosophy and radical politics were perceived to enjoy a natural alliance. Jan Golinski and others have shown how science in the last decades of the eighteenth century was a dominant presence in public culture, in both the provinces and the metropolis, and how particularly both the educated middle classes and women were especially addressed by and responsive to this.[13] Presented not as an esoteric knowledge, but as an essential part of a liberal education, as well as an entertainment suitable for 'polite' ladies of fashion and sensibility, recent discoveries and theories were disseminated through innumerable popularising scientific works, available both to purchase and in circulating libraries; public scientific lectures, such as those given by Joseph Priestley and Humphry Davy, as well as a host of lesser-known figures, also contributed to the 'extension of scientific knowledge through society at large'.[14] For Golinski, such activities were an essential part of Enlightenment public culture, all the more so because, for such a leading figure as Priestley, his identity as an experimental natural philosopher was not wholly divorced from his work as a 'metaphysician, theologian, political writer, and historian'.[15] The anachronistic word 'science', used here retrospectively to characterise what from a distance appears as the disciplinary specificity of these activities, should not prevent us from recognising the close interrelations between diverse Enlightenment practices of knowledge, a closeness which of course aided the dissemination of ideas between different areas of intellectual interest. Such dissemination was also aided by the eighteenth-century clubs, such as the 'Club of Honest Whigs' attended by both Priestley and Mary Wollstonecraft's early mentor Richard Price, or by other forms of social mixing: Joseph Johnson, for instance, the radical publisher of both literary and scientific works (including by Wollstonecraft, Priestley and Darwin), and the radical periodical the *Analytical Review*, was at the centre of a circle of reformist and radical sympathisers, thinkers and writers whose frequent social contact promoted and ensured their mutual intellectual exchange.

As such social networks suggest, a natural allegiance existed in the last decades of the eighteenth century between natural philosophy, Enlightenment reason and progressive politics. At times, this was explicit, even confrontational, as in Priestley's famous declaration that 'the English hierarchy (if there is anything unsound in its constitution) has equal reason to tremble even at an air pump, or an electrical machine', or in Darwin's contention that 'by inducing the world to think and reason', experimental philosophy will 'silently marshall mankind against delusion, and ... overrun the empire of superstition'.[16] (As we saw in Chapter 4, Thelwall had made similar observations in 1793, when the materialist tendencies of his paper 'On the Origin of Sensation' led to his being voted out of the Physical Society: 'from the language and earnestness exerted upon the occasion, one would have thought that the existence of theological and political institutions had depended upon the agitation of a question of physics'.[17]) Such sentiments were met in turn by a suspicion of and opposition to natural philosophy by such figures as Edmund Burke, who used chemical metaphors of noxious gases to represent the effect of the French Revolution, and who resisted the urge to analyse and reduce society into its constituent parts as a sterile philosophical endeavour similar to that of the new science.[18] Inherent in all these sentiments, both supportive of and resistant to natural philosophy, is a perception that science represents not simply a form of knowledge, but a specific mode of thinking and writing – of pursuing reasoned investigations into 'things as they are', favouring the probable and likely, rejecting the outlandish or superstitious, analysing effects into determining causes, and tracing chains of events from outset to outcome – and this in turn points to another affinity between scientific and other, literary forms of writing which equally favour the construction of probable, realist narratives of historical causation. Simon Schaffer and Steven Shapin have demonstrated how natural philosophy depended on certain rhetorical and textual strategies for the acceptance and promotion of its discoveries, and Golinski has shown how, as well as figuring their lectures as refined, polite entertainment, scientific lecturers often presented nature via an aesthetic of the sublime.[19] The particular version of literary realism embodied in the late eighteenth-century woman's novel, with its foregrounding of narrative, its hinging on resonant and likely detail, its persuasiveness, credibility and analysis of chains of historical cause and effect, from this perspective appears to share the same desire to construct analytical, explanatory narrative with contemporary scientific writing.

One location where vitalist science regularly rubbed shoulders with literary, political, moral and economic concerns was the periodical press. In the politically reformist *Monthly Review*, for instance, articles on or related to vitalist physiology made regular appearances in the 20 years from 1780, nestling alongside poems, reviews of moral essays and novels, discussions of Indian or other foreign affairs, and so on. In 1791, Wollstonecraft's *Vindication of the Rights of Men* is reviewed in the same volume as a discussion of a 'Plain display' of animal magnetism, and in 1795, the same periodical's review of Wollstonecraft's account of the French Revolution appears with an address on John Hunter's museum of physiological specimens, Hamilton on the recovery of drowned persons, Tattersall's 'Animal doctrines in favour of materialism', Morgan on electricity and Pugh on muscular motion. Vitalist science in fact appears in a wide variety of different contexts in these pages. The majority of these are straightforwardly scientific: discussions of animal or human physiology, reviews of medical or chemical works, or commentary on the debate between Priestley and Lavoisier on 'vital air' or oxygen, whose discovery added a further dimension to vitality debates. Works by Continental vitalists, such as Barthez, Hufeland or Blumenbach, are well represented and given space for lengthy reviews, and these theoretical works are matched by discussions of the applied uses of vitalist theory: the treatment of scurvy, accounts of Caesarean operations, as well as the recovery of the drowned, all draw in some way on vitalist science. Discussions of electricity, mesmerism and magnetism might also include some reference to vitalist theory, as does James Hutton's geology with its dependence on notions of animal heat. Vitality also makes an appearance in more diverse contexts: in discussions of ancient metaphysics, of the existence of a 'vital principle' in animals, in reviews of Horsley's sermon on vitality or of Warburton's characterisation of faith as Christianity's vital principle, or in theological essays on the immateriality of the soul. In these latter examples, a 'vital principle' segues from its specific scientific meaning to something more metaphorical, but this in itself is an indication of the extent to which an initially physiological terminology, and its associated debate, is being received by and disseminated into a larger culture. Indeed, as Chapter 4's discussion of Horsley's sermon suggested, one reading of such religious responses is that Christianity was having to reaffirm its own account of the origins of human vitality against an alternative, scientific version. This of course was precisely the flashpoint which generated the vitalist controversy of 1814: William Lawrence's claim that the principle of life 'superadded' to matter must itself be material prompted such outcries that he was eventually forced to retract.

Other articles provide further evidence of vitalism's figurative life in the periodical press. At the most conventional, a review of Michell's *Principles of Legislation* suggests that the law of primogeniture, like a vital principle, pervades the whole system of jurisprudence, legislature and manners, to guarantee its strength; or a review of Samuel Stanhope Smith's *Divine Goodness to the United States*, quotes his figure of good government as, like the heart, diffusing 'the vital principle' through every member state.[20] The apparently commonplace nature of such usages itself suggests the way the phrase has entered everyday speech. In these examples, the figurative potential of a discourse of vitality is relatively unexploited, but the same cannot be said of a review of Cogan's *Journey down the Rhine*, which retells his story of a communion wafer endowed with a vital principle through consecration, and cutting itself up into the form of a crucifix.[21] Superstition, religious mysticism and the animation of the inanimate here all appear licensed by the fictional possibilities of a vitalist science which is itself little more than a worker of miracles. But, given vitalism's associations with the more radical elements of what to a conservative establishment already appears a questionable natural philosophy, perhaps the oddest deployment of figures of vitality appears in *The Anti-Jacobin Review and Magazine*. Staunchly loyalist, and launched to oppose a tide of radical sentiment, its first issue opens with a prefatory address which calls unspecified – but no doubt Tory – religious and political principles the 'vital principle' of the body politic.[22] This, as much as anything, is perhaps indicative of the ways a language of vitality and a 'vital principle' detached itself from the relatively constrained, and controversial, scientific discourse on physiology to take on a figurative life of its own in the speech of the time.

Such surveys of periodicals do much to suggest a language of vitality is used across a spectrum of contexts in the last decade or so of the eighteenth century – from precise medical and scientific usages to more figurative or commonplace instances. There is no evidence to suggest that either Fenwick or Wollstonecraft engaged more closely with vitalist debates than might be expected of any general reader of such sources, although Wollstonecraft, in her role as reviewer for another progressive periodical, Joseph Johnson's *Analytical Review*, did review the Scot William Smellie's *Philosophy of Natural History* (1790), a work which makes frequent reference to vital principles, as well as to John Hunter's research, in its survey of plant and animal life. Smellie appears to favour Adair Crawford's theory that breathing imparts the 'vital principle' to animal blood, and somehow unites it with the innate 'phlogiston' or 'fire' in all animals to produce the heat necessary to life. Such a theory does not attract comment from Wollstonecraft in her review, suggesting that it is a commonplace, and indeed phlogiston, as we shall see, informs the imagery of her later *Wrongs of Woman*. Wollstonecraft's knowledge of current science is also implied by a strong attack on Smellie's rejection of the Linnaean system of plant classification. Smellie's account of the 'philosophic idea of organic matter, and the various instincts which modify it' is praised in more general terms, although Wollstonecraft upholds the distinction between reason and instinct, and that of man and beast on which for her it depends, which Smellie had been inclined to blur.[23]

Even such a brief contextual survey as this does enough to show how contemporary natural philosophy played an integral part in late eighteenth-century public and intellectual culture, where it was not siphoned off as an obscure specialism but available to all participants in the Enlightenment republic of letters. The existence of a developed and complex debate into the nature of animation, causes of vitality, the origins of life and generation, and the forces which regulate organised life, such as was constituted by vitalist science, surely makes a powerful case for a context from which to consider the fascination with figures of natural life, vitality and animation which have come to be regarded as constitutive of Romantic writing.[24] But the readings of the texts which follow, however, seek not to show how the language and figures used by these women writers might be brought back to such a context, but rather the reverse: how women's writing between 1789 and 1799 is informed by, borrows and reworks a contemporary cultural fascination with unanswered and, for the time, unanswerable questions pertaining to organic life-forces and vital animation. Of interest here is less how far such language might be mapped back to an origin in natural philosophical discourse, than how it produces certain effects in fiction – in the modified gothic of Fenwick's concern with the conjunction of life and death, or in Wollstonecraft's particular characterisation of genius or sensibility. Such a context helps to explain, I suggest, not only these texts' peculiar insistence on the power and significance of female lives and their productions (imaginative, creative, literary and reproductive), but also their curious counterpull towards death – a marked feature of all the texts considered in this chapter, including, as we have already seen, Radcliffe's story of Fidelia. But if such a context contributes to our sense of the historical moment of these texts, and their participation in larger cultural concerns, than a more confined literary exegesis might acknowledge, they are equally not to be seen as merely slotting into a history of ideas, or being reduced to an explicating context.[25] Rather, it is hoped that what follows illuminates something of the processes by which contemporary influences, contexts, sources and ideas are transformed, refigured and moulded by creative, fictional writers into something which – in that favourite Romantic act (repeated in grotesque form in Shelley's *Frankenstein*) – transcends its own generative origins. If the writers considered below have looser links to the specific, identifiable discourse of vitalism than Thelwall or Darwin, that is because one of this chapter's concerns is not simply to highlight the importance of natural philosophy as an overlooked context for women writers of this time, but to explore how literary culture at a further remove than those immediately involved with it inherited and reworked the scientific preoccupations which, although they have been little acknowledged as informing women's fiction in the decade before Shelley's *Frankenstein*, nevertheless worked their influences in some unexpected, and unexplored, ways.

Fenwick's *Secresy*: gothic vitalism and social criticism

Eliza Fenwick's *Secresy, or The Ruin on the Rock* (1795), is usually discussed by critics – when it is discussed at all – in the context of feminist responses to Rousseau, and contemporary debates about the problems and powers of female sensibility. Its central narrative, of the incarceration of Sibella Valmont by her sociopathic uncle, her forbidden love for her cousin Clement Montgomery, her assertion of independent will and native feeling, and its disastrous consequences (including her pregnancy), certainly invites commentary along such lines. Mr Valmont's segregation of Sibella from what he understands as the malign influences of society is a clear echo of the Rousseauean project of educating women in isolation, and the dependence and absolute obedience he demands from his niece parallel Rousseau's ideals of female passivity and servitude. As Isobel Grundy points out in her introduction to the Broadview edition of the text, Fenwick's friend Mary Wollstonecraft had already accused Rousseau, in her sustained attack on him in *Vindication of the Rights of Woman* (1792), of being a sensualist, a characteristic illustrated too by Valmont in his youth, and Fenwick's tracing of a chain of events which eventually ruin Sibella, Clement and Arthur Murden (a travelling companion of Clement who is introduced to Sibella by her friend and correspondent, Caroline Ashburn), arguably repeats and extends Wollstonecraft's attack on Rousseau by laying the blame for their respective tragedies squarely at the door of Valmont's tyrannical and oppressive power.[26]

Rousseau's account of gender relations also informed contemporary debates on sensibility, a further context in which Fenwick's text has been considered. For Nicola Watson, *Secresy* is best understood as participating in a tradition of female epistolary writing on sensibility newly informed by a debate with Rousseau, but ultimately traceable back to Richardson. For Watson, Fenwick, as with other contemporary novelists, is attempting to validate female feeling as an essential basis for political emancipation, but is 'crippled' by an 'epistolary mode' which returns Sibella to the well-worn fate of the seduced, abandoned, ruined woman.[27] Watson's attribution to the novel's formal narrative mode of such determining powers over plot denouement is all the more striking given her equal emphasis on the discursive conflicts embodied by this epistolary novel's multi-correspondent form. As Julia Wright argues, such a mix of narrators communicates the tensions between alternative outlooks (libertine, rationalist, romantic and so on), and so might be thought to at least map out the possibility of alternative endings for this 'fallen' woman.[28] Critical accounts such as these certainly illuminate something of the literary traditions in which Fenwick was writing, but the abiding impression left by *Secresy* is of an originality, even eccentricity, for which such comparisons with contemporary writing cannot fully account; this sense of the text as a lone, unusual production is underlined by its status as Fenwick's only surviving adult novel. The following pages attempt to offer their own reading of a text which another critic, Terry Castle, found 'incoherent', by tracing the way its themes are informed by a sustained concern with a politicised animation and vitality.[29]

Like that of Mary Robinson's *The Natural Daughter* (1799), Fenwick's plot turns on the existence of an illegitimate child. But where Robinson's text increasingly replaces its initial satire of moral hypocrisy with exhortations to the sentimental virtues of compassionate sensibility and benevolent feeling, a comparison with *Secresy* only reveals the ways in which the latter text resists being similarly read as a novel of sensibility. Neither Sibella nor Caroline, its two most prominent female characters, especially embody the virtues of sensibility, which are instead most readily identifiable in Arthur Murden's early charities to a distressed Indian housekeeper and to the farmer's daughter seduced by his manservant. Rather than pursuing such early themes, however, the uniqueness of *Secresy* is manifested in the ways it becomes caught up in a specific and unusual nexus of language and imagery around the ideas of life, vitality, animation and death – whose compulsion breaks even the putative sentimental hero, Murden. Such notions are especially associated with Sibella herself but also inform the development and plot structure of the work as a whole, which moves from the activity and hopefulness of Caroline's first meeting with Sibella, through the conception of Sibella and Clement's illegitimate child, to the three deaths of Sibella, her baby and Arthur, as well as the banishment of Clement and the metaphorical death, through their deliberate and reasoned suppression, of Caroline's inchoate feelings for Arthur. Turning as it does on the text's central 'secret' animation, Sibella's pregnancy, Fenwick's novel moves beyond the tamer emotional realm of half-expressed sentimental feeling, to explore complex and intertwined themes of life, love and death – ideas linked most obviously in Sibella's meditations, as well as in her fate – and uses them to vary and heighten a critique of Rousseauean philosophy and oppressive social customs which, as other critics have noted, was already well established. It is this insistence on a natural vitality at odds with existing oppressive social laws and practices, rather than on notions of sensibility, feeling and affect, which are most foregrounded in Fenwick's analysis, and the way in which she co-opts such a language and develops its imagery into the feminist novel of ideas, is arguably of most interest in this text. In the process, Fenwick not only enriches the language of contemporary feminist debate, but also develops a distinct literary style, a fascination with imagery of vital life shadowed by death, which might perhaps be termed 'gothic vitalism'. *Secresy* not only produces and exploits a developed language of animation, vitality and their opposites to articulate and extend an existing attack on social tyranny and the oppression of women, but also finds therein both a new means of framing female assertion, newly based on analogy with the observable powers of nature, and a variant on the popular gothic mode, in Fenwick's determined conjunction of life and death.

Fenwick's concern with ideas of vital animation is evident both in the language and imagery of her novel, and in a plot whose central event is the conception of new life – and its death. Where the hero or heroine of the late eighteenth-century novel of sensibility is associated with a lexicon of affect – of feeling, sensitivity, pity, benevolence – *Secresy*'s heroine is given a quite different language, repeatedly associated with the 'glow of animation' visible in her cheeks or the 'fire of vivacity in her eye'. On only his second meeting with her, Murden asks Sibella how she preserves her 'vigour' and 'animation', and Caroline, physically distanced from her correspondent throughout most of the book, is nevertheless warmed by her 'ever vivifying remembrance'.[30] At one level such language simply explains the attractions of the novel's central character, but Sibella's vivacity is also, more importantly, a visible sign of the reasoned self-assertion through which she opposes Valmont's oppressive rule. As she explains, at its root is a capacity for thought, recognised and honed through an education gained from Clement's tutor, Bonneville, which fuels and is manifested in her animation. But thought, the very 'soul of existence', is visible too in the natural world which is the usual location for Sibella's meditations: even the worm thinks, she asserts, even a flower, she states, in a Darwinian observation, has powers of sympathy and reflection. Such assertions suggest the limits of Bonneville's influence in tutoring Sibella; whilst Bonneville is hailed, on his deathbed, as her intellectual 'father', she is, equally, a 'daughter of nature': the hours spent haunting the woods around Valmont Castle leave a stronger mark on her character and outlook than any traces of Bonneville's teachings.[31] The animated powers of life in vegetable nature thus provide the analogical support for her own assertion of mental powers, prompting her to gloss those powers as undeniable facts of her existence and enabling her to oppose Valmont's dictat, 'You shall not reason', as perversely unnatural, an arbitrary social law at odds with those of nature.[32] In her own answer to the questioning of life's origin and meaning which, as the epigraphs at the outset of this chapter show, recur through women's writing of this period, Sibella finds her identity and purpose in the 'stimulating principle' of thought, the 'vivifying principle of intellectual life' which, through language directly drawn from vitalist science, she understands as having created her. So strongly is Sibella associated with the animating powers of life, indeed, that it is death, not life, which prompts her confused questioning, as, in an inversion of the vitalists' usual enquiry into the cause of life, Bonneville's death prompts her to ask after the 'cause' which 'can produce an effect so overwhelming'.[33] Her experience of the sublime fascination exerted by death – 'new, hideous, awfully mysterious' and prompting '[m]ultitudes of dark perplexing ideas' – is one of the first moments when Sibella's hitherto unmodulated vitality begins to incorporate something altogether blacker and more menacing.

This exposition of Sibella's animation, here focused on her reason, develops in a series of interlinked images, which both encompasses established Romantic tropes and extends them in new directions, increasingly to emphasise a coincidence of life and death. The wood where Sibella escapes the confines of Valmont Castle to study animated nature becomes a place of deathly shadows and portents, her own wanderings a 'haunting' of them. The natural shades into the supernatural too in the representation of Sibella's capacity for thought as Promethean, a near-divinity implied in Caroline's acknowledgement of Sibella's ability to transcend her own more solidly quotidian acts of reason, or, more complexly, in the libertine Lord

Filmar's transformative recognition of Sibella as an 'angel'. More explicitly, such qualities distinguish her from the Valmont who lacks the ability to steal fire from heaven, a comparison which anticipates that between Maria and her husband in Wollstonecraft's *Wrongs of Woman*; it also differentiates Sibella from the 'inanimate' Mrs Valmont, her uncle's wife, so lifeless that her tomb is already being built, or from those, including Caroline's mother, Mrs Ashburn, whose pursuit of 'mechanical pleasures' turns her into an automaton.[34] Such contrasts between the vital female 'genius' and the lifeless 'lady' are far from unique to Fenwick (they appear too, of course, in Wollstonecraft's early novel *Mary* and her *Vindication of the Rights of Woman*, and are still echoing in Austen's Lady Bertram in *Mansfield Park*), but Fenwick pursues the implications of such relatively conventional images into the very unfolding of her plot. Sibella's heightened vivacity at times renders her, paradoxically, otherworldly – her extreme embodiment of apparent suprahuman animation setting her apart from others almost as a different kind of being, and rendering her eventual death interpretable as the inevitable demise of one not quite of this world. There is a kind of deathliness to her, it emerges, despite, or perhaps because of, her vital animation. It is this sense which echoes in Filmar's awed address not just to the 'angel' but also to Sibella the 'spectre', or in the quasi-medieval formalities which Murden chooses to use when he encounters her in the wood near Valmont Castle, or indeed in the whole extended apparatus of gothic imagery and machinery which first unfolds Sibella's story, and later entraps her.[35] It is precisely this conjunctive embodiment of the extremes of life and death which fascinates Fenwick and eventually brings to an unfruitful, tragic end the promising assertions of the natural life-force of reason with which her novel opens.

These combined extremes of life and death are brought into a more pressured conjunction in Sibella's extension of her animating principle, her vital power, to include not only thought and reason, but also love. His love for her, she believes, means that she is Clement's 'actuating principle' or 'soul', and love's status as the animating engine of life is such that 'our hearts must cease to throb with life, when ... love is extinguished'.[36] Where the vital powers of reason appear strengthened by such analogies, however, their extension to include fervent expressions of romantic love begins to reveal their limitations, and to make a language of vitality a sign of overblown idealism: no longer an assertion of innate critical powers, it becomes a symptom of the effects of Valmont's dangerous isolation of her. This is especially emphasised by Caroline's reasoned repression of her incipient feelings for Murden, and by Murden's recognition of his own idealising love for Sibella as a 'disease of the mind'. Sibella's persistent framing of love as equivalent to life, as a necessary vital force, when reiterated alongside the accumulating evidence of Clement's unfaithfulness, predicts, even precipitates, her eventual demise when the extent of his betrayal of her is revealed.

Her identification of love as a necessary principle of life only renders her more vulnerable to the failings of others, turning their weaknesses and inadequacies into her own death blow. Far from being a means by which the body is transcended, as the 'soul expands into a new existence' and the 'body's encumbering mass seems no longer her organ', as she exults at one point, her insistence on love as an essential source of her animation only reiterates the precisely physical bounds of her existence, when that essential love is withdrawn.[37]

Sibella's identification of love as a vital power which enables the 'body's encumbering mass' to be shed, attempts to refigure vitality into a means of transcending the body. But her pregnancy, of course, is a dramatic statement of the opposite: a visible trace of the physical expression of her love for Clement evident in her body's reproductive powers. Indeed, Sibella's association with animation and life – with a 'vivifying principle' – is most obviously manifested in the text through the conception of her child, and the eventual fate of that child, who is stillborn in the aftermath of Sibella's traumatic discovery of Clement's marriage, operates as a commentary on her sustained attempts to assert the forces of life and love against the various oppressions she endures. Elsewhere in women's writing of the period, too, the child is assigned a powerful signifying function: a device, in Mary Robinson's *The Natural Daughter*, to expose libertine behaviour, sexual immorality and (given its conception in the chaos following the French Revolution) the domestic effects of political upheaval; a sign of possible optimism, as well as a focus for maternal anxiety about female suffering in Wollstonecraft's *Wrongs of Woman* and *Short Residence*. Sibella's child, dead at birth, however, can only offer a negative meaning: an embodiment of the happiness denied to her mother, of love lost and life extinguished. Given Valmont's ultimate responsibility for much of what takes place in the text – his refusal to sanction Clement and Sibella's relationship, despite his complicity in their mutual affection, his manipulative treatment of both – Sibella's child also signals the deathly, fruitless outcome of his various tyrannies. The unnamed child is thus, like Frankenstein's later creation, a kind of malformation (hence perhaps the horror with which Sibella's pregnancy is greeted by all but the rakish Filmar): rendered such by the social laws which refuse to acknowledge the relationship which has given it birth, as well as by the delusions, ignorances and constraints of its mother and father, whose abortive love it ultimately signifies.

If at the outset of this novel Sibella asserts that thought is her 'vivifying principle', by the end, what has been animated for her is a more material version of life which returns her to, at best, the traditional fate of her gender, mothering, or at worst, the equally familiar death of the seduced, fallen woman. What Fenwick's novel brings to life is thus from one perspective a very old story about gender roles and relations, even if it is one which hopes to offer social criticism in the midst of its tragic runes. Considered in this way, Sibella's association with the powers of vitality is deflected onto a familiar tale of female fates, and Fenwick's exploitation of a discourse of animation appears defeated by a seduction tale which at least narrates also the oppressive nature of established social custom and gender roles. Equally, however, the effect of Fenwick's emphasis on the powers of animated nature, on vitality and vivification, is to underline the deadly effects of human error, tyranny and oppression, and to offer a powerful contrast between the natural forces of life, and the human acts, laws and customs which repress them. In this, her version of animation, at critical odds with established social custom, operates in clear contrast with the rather different religious animation which saves that other fallen woman, Fidelia, at the conclusion of Radcliffe's *Female Advocate*. Where Fidelia's version of animation works to return the submissive, compliant, humbled woman into a suitably lowly place in the social order, Sibella's faith in the animating powers of reason and love renders her an icon in her death of what has been lost, and a sign of the failings and dangers of Valmont's unsustainable ideals. In *Secresy*, a discourse of animation traceable back – through its Darwinian perceptions of animated nature, and its fascination with vital forces and animating causes – to contemporary vitalism, is both harnessed to the cause of female emancipation and made to be a forceful participant in current political debates. The critical force of a vitalism controversial in scientific circles for its materialist challenges to a God-created world is transferred in Fenwick's novel to inform a parallel challenge to socially oppressive laws and customs which Sibella's demise, and that of her child, only makes appear all the more deathly.

Mary: genius and improvement

Mary Wollstonecraft's early novella *Mary* (1788), equally concerned with female suffering, oppression and happiness, offers some obvious similarities with the later *Secresy*. Although Fenwick and Wollstonecraft were friends (Fenwick nursed Wollstonecraft during her final illness after the birth of the future Mary Shelley, and took care of the baby in the immediate aftermath of Wollstonecraft's death), the texts have rarely been considered alongside each other. But *Mary* and *Secresy* arguably have much in common: a shared focus on the struggles for self-assertion and liberty of a central heroine constrained by defined gender roles and the schemes of her guardian, an identification of female friendship as a rare source of personal expression, and a passionate pursuit of interdicted romantic relationships – themselves regarded, finally (but perhaps mistakenly), as the ultimate locus of personal fulfilment. By the time she writes *The Wrongs of Woman*, as Claudia Johnson points out, Wollstonecraft is able to move the plot of female emancipation beyond the idealised and sentimental heterosexual romance, to a new vision of female solidarity and communality based on a recognition of shared oppressions, but both *Mary* and *Secresy* use their most extreme and elevated language to tie the fates and happiness of their heroines to the success or otherwise of their romantic attachments.[38] Mary's 'I cannot live without loving, and love leads to madness' anticipates Sibella's uncompromised linking of love and life, and, whilst Mary survives Henry's death, there is nevertheless a strong counterpull towards the death which is more climactic in Fenwick's text, both in her earlier desire to share a 'watery grave' with Henry, and in the novel's notorious final sentence, which looks to death as a place of negative fulfilment, where '*there is neither marrying*, nor giving in marriage'.[39]

The language of animation used by Mary to describe her most essential being provides a further point of connection between these texts. Where thought, and later love, are identified by Sibella as her 'vivifying principles', for Mary, as she tells Henry, 'involuntary affections', fixed by reflection, are 'interwoven' in her soul and 'animate' her actions.[40] Where *Secresy* pursues its themes through a sustainedly foregrounded language of animation, however, *Mary* offers an ostensibly alternative focus in the 'genius' which is highlighted in the text's epigraph, taken from Rousseau's *La Nouvelle Hèllöise*: '*L'exercise des plus sublimes vertus élève et nourrit le génie*'; the fate of vitality in this novel will be closely linked to that of genius. For Wollstonecraft, the interest of 'genius' appears to have resided in what the verbs of this deceptively simple phrase emphasise – the activities of improving, elevating, nourishing and refining an original power of genius. This certainly is the sense in which she responded to another text on this theme, John Weddell Parson's *Hints on Producing Genius*, which she reviewed in the *Analytical Review*. Noting that genius has 'generally been considered as an arbitrary gift of nature; that is, an individual character capable of improvement, but not of alteration', she praises Parson's recommendations to strengthen the body and exercise the mind, and concludes that, if these hints were followed, 'the race, improved gradually, during many successive generations, might all be men of genius, compared with the present dwarfish, half-formed beings, who crawl discontented between earth and heaven'.[41] For Wollstonecraft, genius is closely linked to an Enlightenment creed of improvement, both personal and for all mankind. From this perspective, its epigraph invites us to read *Mary* as precisely such a tale of genius tamed, honed and improved – or at least an attempt at this. And for a critic like Gary Kelly, such improvement is achieved in the text through a taming of sensibility and imagination via self-education and rational religion, so that *Mary* ultimately offers the expression of an 'authentic self' in critical relation to the society which produced her.[42]

This sense of genius as a gifted individual at odds with society certainly chimes with Rousseau's own conception, but the notion of genius also evokes a whole other debate about the source and nature of the active, original, creative power which genius was supposed to be. For Rousseau, genius's 'celestial flame which warms and sets fire to the soul' implied the same Promethean origins which Wollstonecraft (and other Romantic writers) would associate also with the imagination, but in other contexts the question of genius's origins was less easily solved.[43] In eighteenth-century aesthetic debates, genius emerges as an original, natural power stemming from unknown sources, a definition which leads to its characterisation in organic terms, especially when contrasted with more mechanical or imitative talents or productions, acquired more prosaically through learning or rote application. Such an opposition is expressed in striking terms in Edward Young's *Conjectures on Original Genius* (1759), an author quoted by Wollstonecraft both in *Mary* and elsewhere. Distinguishing between an original creative production and an imitation, Young suggests that the former may be 'said to be of a vegetable nature', rising 'spontaneously from the vital root of genius', whilst imitations are wrought by 'mechanics, art, and labour'.[44] Young's characterisation of genius as stemming from, or having, a vital active power, looks to natural philosophy, where innate natural powers had been debated since Newtonian times, and where, by the Romantic period, the investigation of such active powers was, as Simon Schaffer has claimed, natural philosophy's 'most characteristic aspect'.[45] Schaffer lists electricity, pneumatics and galvanism as examples of Romantic natural philosophy's pursuit of active natural powers, but of course vitalist physiology's attempt to attribute and theorise active powers of life within animal and human bodies is also pertinent here. Young's language foregrounds the way in which aestheticians' attempts to comprehend the original creative powers of genius parallels natural philosophy's enquiry into the original organic powers of life. Wollstonecraft's exploration of genius, as thematising analogous unknown, active, original creative powers, thus reiterates, in a different vein, a persistent enquiry of contemporary natural philosophy, but as *Mary* unfolds, the difficulty of finding a secure language in which such powers might be expressed and manifested becomes clear. As the following discussion suggests, this difficulty in representing the 'vital active power' of genius also indicates the challenges of understanding and theorising such a power's exact nature and operation – a challenge which is finally only overcome in the last pages not of *Mary*, but of Wollstonecraft's later *Wrongs of Woman*.

One of the first obstacles negotiated by *Mary* is the need to trace the historical unfolding of something whose origins are by definition always unknown. The problem is already encapsulated in *Mary*'s epigraph, which looks in two directions: on the one hand it invokes the original active power of 'vital' genius, a power whose origins are unknown, and which is not taught or mechanically acquired; on the other, it also insists on the possibility of refining and improving that which cannot be instrumentally generated or produced. As a consequence, the text's account of Mary's genius is located in the difficult space between origin and being, between existence and self-becoming, and this perhaps helps to explain why its narrative structure, whilst emphasising historical causation and biographical development in its account of Mary's life, also runs the risk of being slightly 'after the event', as understanding the processes of Mary's improvement will never answer the great unknown of how her 'genius' came into existence in the first place. Equally problematically, the same narrative commitment to notions of cause and origin – demonstrated, for instance, in the way the text begins not with Mary herself but with her mother – can produce some strange causational effects: Mary's emerging sensitive and inward character, for instance, can appear a consequence of her emotional neglect by her dysfunctional family, her formative philosophical meditations on life and its passing, a morbid consequence of the traumatic witnessing of shocking deaths at an early, impressionable age. A historically predicated narrative cannot help but assign causes even to what is ultimately unknowable, and genius here, far from being a mysterious visitation of divine powers, risks being assigned all too earthly an origin in the ordinary failings and minor tragedies of domestic, familial existence.

The narrative commitment to historical unfolding begins to seem even more problematic as what is unfolded emerges less as the 'elevation' and 'nourishing' of genius announced by the epigraph, and more as the progression through accumulated unhappinesses and constraints to the death-wish of the denouement's eventual stalemate. As has been noted, Kelly secures a more positive reading of the text by emphasising how the sufferings of self-developed genius expose the critical failings of society, but such a reading depends on understanding Mary as having achieved a mature honing of her character, which more attention to what is unresolved and fragmented in the text might throw into question. Wollstonecraft's own late evaluation of the novel as a 'crude production' suggests her own sense of its limitations.[46] In fact, *Mary*'s inability to formulate a secure language in which the active power of Mary's genius might be expressed, and to identify a vehicle in which it can be manifested, begins to tell as a central failing; in this, it might be seen to share with late eighteenth-century natural philosophy not only an interest in such original organic powers but also its difficulty in theorising or defining them. Where *Wrongs of Woman* eventually offers a specific account of sensibility as an animating power whose vitality sustains Maria through the various betrayals by her husband and Darnford (if not, finally, beyond some of the imagined deaths which Wollstonecraft postulated as conclusions to the unfinished work), *Mary* explores various alternative modes by which Mary's animating genius might be expressed, including imagination, sensibility, religiosity and love, without settling on one, and this contributes to the sense of a text, and a life, which by its conclusion has reached a kind of dead end. Where, in later Romantic texts, the failure of vehicles for expressions of genius, imagination and poetry itself has a recuperative value in underlining the sublime ability of the poet to transcend them anyway, in *Mary*, similar failure leaves her not elevated but stranded in a loveless life.

Mary's difficulty in formulating a mode of vital animation for its heroine is most noticeable in the relatively limited attention given to imagination in the text. This is especially striking not only in contrast with the later *Wrongs of Woman*, where imagination plays a full and crucial role, but also given the existence of a well-established theorisation of imagination and creativity in aesthetic discourse well before 1787, when Wollstonecraft was writing. Mary's rhapsodies only occasionally address the powers of imagination, however, and when she identifies what Sibella would call her 'vivifying principle' in conversation with Henry, she foregrounds her feelings: love and affection are primarily what animate her. Where Wollstonecraft elsewhere (conventionally enough) identified imagination as the Promethean animating fire, linking the earthly and the divine, Mary quotes Milton to suggest that 'earthly love is the scale by which to heaven we may ascend'.[47] Love, which she asserts will be unchanged even in death, is the primary principle of transcendence, but such ideals soon prove problematic in relation to both Mary's charitable and romantic impulses. Her various acts of charity clearly express a benevolent nature capable of sympathy and pity, but such 'exercising' of her soul, understanding and imagination, often – as with her attentions to the woman rescued at sea – only problematically highlight a qualitative difference between Mary's more animated nature and the objects of her actions.[48] If Mary is thereby ascending the scale to heaven, she is clearly leaving others very definitely beneath her. Sensibility, manifested here and elsewhere, also ultimately proves unable to sustain Mary against the melancholic resignation to which she eventually capitulates; failing to generate a more satisfactory resolution to her difficulties, it appears only to sensitise her to her sufferings, and to give her a sense of difference from others, rather than the communality in which Maria, as we shall see, exults. It emerges finally, perhaps, as simply an unhelpfully refined expression of discontent and oppression: valuable no doubt in a text whose purpose is understood as the recognition and exposure of the social ills by which women suffer, but less convincing as evidence of genius 'nourished' and 'elevated'.

Mary's sensibility is closely allied to her religiosity, as she asserts in one of the rhapsodic meditations which erupt through the text, and the language she uses in relation to her faith promises to align it to something like a motivating active power: it is the 'chimera' of the 'medicine of life', and virtue should be life's 'active principle'.[49] The fragmentary nature of such effusions, however, perhaps already classes them as inchoate, and their potentially energising effects are soon swamped by a competing sense that religion's real worth is to teach the value of resignation to life's inevitable hardships and sufferings. The rhapsodic religious effusions of the early parts of the text have by the end given way to this later stoic endurance, which looks to death as a place of ultimate release. Such emphasis on religion as the arena in which the major questions of Mary's life might be answered makes Barbara Taylor's assertion that *Mary* is above all a search for 'authentic religious subjectivity' convincing, especially when compared to Wollstonecraft's own contemporaneous formulations.[50] In a letter written in 1787, the year of *Mary*'s composition, Wollstonecraft asks why 'have we implanted in us an irresistible desire to think – if thinking is not in some measure necessary to make us wise unto salvation'.[51] Thinking, the act which for Sibella had simply been sufficient evidence of her animation and existence, is for Wollstonecraft intimately linked to religious eschatology: questions of life cannot be posed without a framing sense of religious ends. Christian faith, finally, is all that remains to Mary, with the loss of Henry and the romantic love which for a while figures as the most promising vehicle for the higher impulses of her genius, but far from being a means of elevation, it offers a hard lesson in submission to the sufferings of life before which all are equal.

In *Mary*, a text which announces its intention to understand the processes by which genius is generated and improved, finally fails to offer an account of a vital spirit in its heroine capable of withstanding the amassed sufferings and obstacles she encounters as her story progresses. Despite the narrative commitment to developmental unfolding, the text's ending, looking forward only to death, speaks to the generative failure of what has been nurtured in Mary. Unlike the plants in Erasmus Darwin's botanic garden, so precisely adapted to flourish in their respective environments and especially in their love lives, Mary and Sibella's difficult lives reveal an animating spirit at odds with society and which in neither case can ensure their happy and prosperous survival. But if *Mary* finds it difficult to envisage the flourishing and improvement of genius, it is a text which nevertheless asserts women's animated existence, even in the face of its own failure, as evidence of the genius of its author. Like Wollstonecraft herself, in her letter of 1787, asserting her powers of thought, Mary is convinced that she is '"*of subtiler essence than the trodden clod*"': an assertion of faith in futurity but also of vital life in the present.[52] If such assertions from Mary ultimately generate little for her, the existence of the text itself, with the claim to female genius which it embodies, nevertheless constitutes a bolder declaration: an assertion of the female genius, and creative imagination, which was considered impossible by Rousseau and others. Against Rousseau's claim that women's writing never contained 'soul', or William Duff's opinion that women lack 'creative power and energy of imagination', Wollstonecraft's 'Advertisement' which prefaces the text uncompromisingly insists on her novella as animated by generative authorial feeling: a double assertion not only of her own genius, but its manifestation in her authorial creation.[53] Making use of the same distinction between copies and original which informed Young's account of genius as a 'vital' power, she dismisses the art which merely 'copies', and claims as her own that in which the 'subtile spirit of nature' is caught. The language of vital animation briefly used by Young is extended in a series of complex statements linking authorial animation to that of her work: it is the 'soul' of the author which should 'animate the hidden springs' of a fictional work, and the scenes created should be varied by a 'vivifying principle'. Something more than the 'gross parts' of nature are presented here: fiction allows us to see the real existence of 'the mind of a woman'. Just as her later novel would claim to 'embody' the sentiments of women, here Wollstonecraft makes clear that her first fictional work displays the powers of the female mind to create, animate, sustain; more than representational art, it is itself a calling into existence, a vital creation. If genius finally cannot be represented, or its origins uncovered, its effects can nevertheless be displayed. If the origins of genius's vital powers cannot be known, if a language in which it may be represented has not yet been found, Mary's animation, under stress within the confines of the tale, nevertheless speaks to the vital creative powers of her maker.

The Wrongs of Woman: from improvement to animated sensibility

Mary's preoccupations with genius, sensibility, imagination and love are picked up again and reworked in Wollstonecraft's final, unfinished novel, *The Wrongs of Woman* (1798), where the earlier text's search for a power capable of sustaining its heroine's fraught energies finds a possible answer in a recognition of the vital powers of sensibility. With its similarly named heroine, Maria, this later text also addresses what it terms the 'partial laws and customs of society' through a focus on a representative central figure, whose qualities and resources of character both highlight the problem and present possible solutions to the 'misery and oppression, peculiar to women'.[54] This particular emphasis on the psychology, affections and inner experiences of a character, a consequence of the novel form despite Wollstonecraft's insistence that the story 'ought rather to be considered, as of woman, than of an individual', means that for extended sections of the text dramatic action is suspended whilst Maria's extended cogitations and reveries are narrated and considered. Although, in recall, *The Wrongs of Woman* can appear to the reader a more dramatic book than its precursor, *Mary*, the majority of its action either takes place in the final quarter of the book, or is related retrospectively in the three inset narratives of Maria, Jemima and Darnford. The principle of historical causation and biographical development which had structured Mary's narrative, whilst still in some sense in place, is thus modified and periodically suspended by a story which begins *in medias res*, and which looks back to the events which have brought its characters to where they are in the present as much as forward to their eventual futures. In this way, the commitment to historical cause, origin and development which constrains the earlier text is alleviated, bringing a greater formal freedom which is echoed in Maria's more emancipated, if still unresolved, experiences.

The Wrongs of Woman's looser narrative structure allows the meditative, near-discursive concerns which preoccupy Maria in the asylum to take centre stage as the novel opens, as Maria's physical incarceration dictates the determined stasis of the text's first sections. Mind and the improvement of mind, the recurring objects of Maria's thoughts, are thus constantly offered for the reader's attention also, emerging as the dominant themes of these opening chapters as Maria concludes that her best chance of securing her escape depends largely on the state and resources of her own 'active' mind. The madhouse setting enables ready distinctions to be drawn between healthy and diseased minds, and observations to be made on the dangers of the passions, or on sickly or strong imaginations; Maria's own creative musings on the mind which is revealed in the marginalia of books borrowed from another inhabitant of the asylum further develops these themes of the dangers, as well as promises, of fancy and imagination. Qualities of mind, rather than improvement of character, are thus foregrounded in this text's exploration of the vital, creative powers needed by women in their negotiation of social oppression.

The theme of imagination in this text – its seductions, possibilities and dangers – has of course been well noted, but it exists in this novel within a larger preoccupation with the 'improved' mind which recalls similar concerns in *A Vindication of the Rights of Woman*, as well as in *Mary*'s concern with the improvement of 'genius'. The improvement of the female mind was of course the object of numerous conduct books – a genre to which Wollstonecraft herself contributed with her early *Thoughts on the Education of Daughters* (1787) – and is explicitly incorporated into *The Wrongs of Woman* with Maria's letter of advice to her daughter, in which she states that 'your improvement' is 'ever present to me while I write'. Such a structure, of a familiar letter from mother to daughter, was a favourite device for many conduct books, including, for example, Lady Sarah Pennington's popular *An Unfortunate Mother's Advice to her Absent Daughters* (1761).[55] *The Wrongs of Woman* ultimately marks its difference from these other texts by transforming the theme of improvement via a language of animation such as was sought in *Mary* but never successfully maintained. As we shall see, Maria's 'improvement' in *Wrongs of Woman* is eventually achieved by securing a language of animated mind and being which ultimately transcends the text's advice book parallels to transform conduct literature's polite and constrained expectation of 'improvement' into a bold claim of a higher nature equally unshackled by earthly constraints and social customs. Offering what is effectively a transformed perception of 'mind', it in turn produces new readings of imagination and sensibility – already identified as key terms in *Mary*, but again never properly developed – as animating, creating forces in which lie the answers to the dilemmas faced by the 'woman' named in the title and author's preface. Where the final sentence of *Mary* offers the bleak prospect of death understood as offering redemption through a series of negatives ('*neither marrying*, nor giving in marriage'), *The Wrongs of Woman*, in at least one of its projected denouements, ends by insisting on the recuperative, reanimating powers of rediscovered motherhood.

The theme of improvement comes to the fore when, at the end of Jemima's long and varied autobiographical narrative, Maria and Darnford comment on the difficulty of improving the mind when living in poverty, and on how the affections especially need regulation by an improved mind. As an interpretative pointer, and although in part a social criticism, such comments invite a reading of Jemima's narrative as a continuation of the thematic lines already opened by Maria's own musings on mind – in addition to the other obvious themes suggested by Jemima's various and appalling experiences: the sexual double standard, the difficulties of women's economic survival, the effects of maternal neglect. Jemima's 'improvement', not simply of mind but in her humanity, is of course the trajectory of the larger narrative in which she appears: her relationship with Maria traces the thawing of one 'benumbed' with near 'petrified ... life's-blood of humanity', 'blighted' at 'the very threshold of existence', to her own act of reviving Maria in the final postulated conclusion to the text. Such language, of the freezing of vital or organic forces of humanity, is repeated in Maria's fears about the '"killing frost"' to which her daughter's 'tender blossom' would be exposed.[56] Benighted improvement is highlighted too in the author's Preface, where Godwin, as editor, includes an extract from one of Wollstonecraft's letters which expounds on how marriage to a man of lesser capacity degrades a woman of 'improving' mind, preventing the cultivation of taste and the fostering of love. Such a process is of course depicted exactly in Maria's own biography, where her taste is 'cultivated' and faculties 'ripened' by married life in London, only for the 'energies of her soul' to wither to the point where only the company of a casual visitor reminds her that she had some 'dormant animation, and sentiments above the dust in which I had been grovelling'.[57]

It is at this point that the language of improvement develops from the static formulations of the advice book, with its horticultural overtones – the cultivation of taste, refinement of sensibility and so on – to a more dramatic, even hyperbolic realm, so that improvement is less a simple growth or pruning, more a participation in a different order of existence, a higher level of animation. Such phrases might recall Mary's earlier awareness of an essence more subtle than 'the trodden clod', but whilst her formulation refracts a religious sense of bridging a difference between the mortal and the divine, Maria's language reflects a less specific, broader appeal to the powers of the living, to an animated nature in which she includes herself. Maria's willingness to seek and embrace the transcendent difference of fully animated being is articulated most fully in her comments on leaving her husband. The sense of distinction she feels in comparison with him is more than that of a difference between species – 'as if an ape had claimed kindred with me', a

line of thought which occurs to Mary, too – than the sense of belonging to a
different order of being, a higher class of animation. 'I was ready to imagine
that I was rising above the thick atmosphere of the earth; or I felt, as wearied
souls might be supposed to feel on entering another state of existence.'[58]
Such moments are felt too, fleetingly, by Mary, but are often sustained only
as the troubling recognition of material difference between her own state
and that of the poor – a recognition which prompts the acts of charity
which only heighten and continue that sense. Maria's more metaphysical
articulation of the same sentiments, unclogged by the urge to smooth what
Mary understands as the material manifestations of differences of nature,
rises more readily to a transcendent and sublime sense of self which, lacking too Mary's turn to religious effusions at moments of heightened self-
awareness, finds a self-sufficiency in her own feelings alone.

Maria's metaphysical language here picks up the theme of animation
which, alongside improvement, has been running through the book, and
which ultimately emerges as a solution to questions of improvement –
a solution which transcends improvement itself. A renewed animation in the
countenance, actions and 'humanity' of Jemima is one of the few positive
achievements of the narrative, but such themes are fleshed out too through
the figure of Maria's husband, George Venables. His lack of the 'fire' needed to
'ferment' the 'faculties', a comment which recalls Valmont's unPromethean
nature, might be thought a character-failing such as might be remedied by
proper attention to self-cultivation, but the diagnosis already looks beyond
such superficial 'tending' of character to an entirely different language. As
the commentary on Venables is developed, the animation or energy which
he lacks emerges as a quality of an altogether different order. For Maria,
Venables is not entirely without 'the generous emotions' which are the
foundation of virtue, but rather they are 'frequently ... so feeble, that, like
the inflammable quality which more or less lurks in all bodies, they often lie
for ever dormant'.[59] This mysterious reference to an innate quality of inflammation (a reference glossed in neither the Oxford nor Penguin editions of
the text) points not to a Promethean fire, but to phlogiston, a fifth element
believed since the ancient Greeks to be contained within all combustible
bodies, and which was released when they burned. Phlogiston theory, whilst
not central to vitalist physiology, was nevertheless a vitalist theory because
of its assumption of a specific property or power in bodies responsible for
the action of inflammation; as Wollstonecraft wrote, its existence was being
challenged by the French chemist Lavoisier, and defended by Priestley who
called the oxygen he discovered 'dephlogisticated' or 'vital' air. With its
implication of an innate internal power responsible for firing character, the
phlogiston reference in Venables' description points to an elemental failure
in him, who thereby is seen to lack an essential power of vital life, the mark
of animated being which has been rediscovered in Jemima. Venables' failing
is all the greater because it is in this animation, the foundation of virtue,

that humanity resides. Eventually resorting to 'wantons of the lowest class' to 'rouse his sluggish spirits', the absence of Venables' vital heat is also linked to sexual impotence – a move which suggests, by way of contrast, that Maria's own sexual self-liberation, in her relationship with Darnford, expresses her own powers of animated self-determination. Elsewhere too, other forms of sexual behaviour suggest false or problematic versions of 'animation' – a word to which Wollstonecraft repeatedly returns. For the libertine philosopher for whom Jemima works as a housekeeper, 'animation' is simply a function of his drinking; but Darnford's sexual profligacy in his early youth, and again in the novel's final stages, indicates that the 'animation' initially visible in both his marginalia and physiognomy is ultimately of a questionable nature.

The language of animation against which Venables and Darnford are measured – and fall short – is also that in which Maria's own improvement is pursued. In an early attribution of the 'electric spark of genius', Maria is shown to have what Venables lacks, but in quite what form this genius takes, and how it will aid her benighted situation, remain to be seen.[60] The term 'genius' immediately recalls Rousseau, who, in the context of a reference to *La Nouvelle Hèlloïse*, is described by Darnford as the 'Prometheus' of sentiment, possessing the 'fire of genius' – a comment which also recalls *Mary*'s framing of its exploration of genius via its epigraph from the same Rousseau text.[61] It is only a few pages before this that genius is attributed to Maria, and thus, in a more integrated, less anxiously assertive way than *Mary*'s epigraph, Maria is shown to share, or surpass, Promethean powers ascribed to the French author. But whereas Rousseau's powers of sentiment are confined to the literary writings in which they are displayed, the animating powers of Maria's genius roam more widely, containing both promise and dangers. Most specifically, it is manifested in the imagination, whose generative powers can be double-edged. Like the mad with whom she shares her asylum, and whose lunacy in part resides in a failure to control their animating powers of thought, Maria's imagination is capable of unrestrainedly embodying 'phantoms of misery'. The dangers of imagination are plotted especially clearly in relation to romance: Maria's conjuring, Pygmalion-like, of ideal qualities to adorn her image of Darnford, repeats a process which had already taken place in her earlier, disastrous attraction to Venables. Irony intercuts the consummation of Maria and Darnford's relationship, with Maria's happiness at finding 'a being of celestial mould' undermined by the way Darnford is presented as an imaginative projection, 'plastic in her impassioned hand', reflecting her sentiments back to her.[62] But at the same time, 'the moments of happiness procured by the imagination, may ... be reckoned among the solid comforts of life': the imagination's fantastical pleasures nevertheless have a kind of substantive reality. The 'genius' of the animating imagination – what it can and cannot create, and the consequences of that creation – is thus clearly central to the questions of female happiness, oppression and fulfilment pursued by the text, even if, as it will turn out, romantic happiness may be rather less essential. Animation, creation and generation in the context of the imagination, hold forth the promise of some solution to female 'misery and oppression', and the possibility of women realising, through them, what they might become.

Maria's qualities of animation, energy and vitality, manifested in the 'genius' of her imagination, are also more generally evident, and contribute to a specific representation of her sensibility as an active energy and animating force. Maria's energy of mind and action are noted from the outset of the book, where her 'extraordinary animation' (reminiscent of Sibella's) is feared by Jemima as a sign of madness; it appears dangerously close to the 'intoxicated sensibility' and 'restless activity of the disturbed imagination' of the genuinely insane.[63] This emphasis on Maria's vitality contrasts with the all-too-familiar representations of feeble femininity and nervous sickliness elsewhere (in Mary's mother, Eliza, or Sibella's Aunt Valmont, for instance) – a sign of the inevitable female obsolescence which, as Andrew McCann has noted, also 'lurks throughout the *Vindication of the Rights of Woman* as the dark fate of the object'.[64] Maria's energy of mind and being, so threatened by her marriage and stimulated by her meeting with Darnford, eventually finds a secure field of expression not in the relatively narrow, and potentially delusory arena of her romantic relationship with her lover, but in the broader sphere of her life more generally, where a newly animated sensibility forges a renewed sense of her sociable connection with others, casts what sensibility itself might be in a new light and gives a new turn to her 'genius'.

Sensibility's links to active virtue and imaginative 'fire' are foregrounded throughout *The Wrongs of Woman* – as when Maria denounces women who sleep with their husbands out of compassion or duty ('they want that fire of the imagination, which produces *active* sensibility'), or when false, affected sensibility is stated to betray not a concern with others but self-regard.[65] It is not *any* sensibility, but a particular version of it, which carries Wollstonecraft's hopes for Maria in this text; Maria's emancipation from the many burdens of sexual and social oppression in the novel depends on the elucidation of Wollstonecraft's particular version of animated, active sensibility. The specificity of this version of sensibility – on which hopes of rescuing Maria are pinned – is even clearer if we recall the sheer variety of ways in which this term was used in the late eighteenth century. As Markman Ellis has usefully reminded us, 'sensibility' could mean many things at this time: a force animating religious feeling; a capacity to be moved by objects and events; a form of aesthetic or moral discrimination; an aspect of refinement and claim to gentility; or, more pejoratively, an assumed mode of fashionable comportment.[66] For Harriet Guest, in another analysis, sensibility's volatility renders it a 'language of fantasy and of sexual and material desire'.[67] Sensibility's links to a physiology of sympathetic response and nervous sensitivity – a physiology closely linked to the vitalist concerns of Whytt, Hunter and others – have been well documented, but Wollstonecraft's development of the term echoes the related discourse of animation, of vital energies and life-forces in contemporary science.[68] Such vital animation is certainly what emerges as the keynote of Maria's sensibility: sensibility as a galvanising fire, an innate internal energy, like phlogiston, which, properly fired up, animates character to virtue and feeling. Originally conceived by Adam Smith, in the most extended philosophical theorisation of moral sensibility, as a power of imaginative intuition – the ability to sense the feelings of others through one's own imaginative sympathy – in Maria, this 'fire' is extended beyond the discrete acts of moral discrimination which Smith describes, to fuel her refutation of an entire system of social hypocrisy and inequality, and to power her self-justifying speech in court. The sensibility which is proclaimed, perhaps a little tamely, as the 'auxiliary of virtue, and the soul of genius', in such more expansive contexts also provides the energy needed for the socially and sexually revolutionary acts which challenge female 'misery and oppression'.[69]

Maria's representations to the judge in court, at the end of the novel, offer the most extended and analytical account of her oppression, but given that her pleas – underpinned by her 'vitally animated' sensibility – fall on deaf ears, the powers of her animated sensibility are perhaps more effectively demonstrated elsewhere. Claudia Johnson has noted the significance of imagery of natural blossoming in *The Wrongs of Woman*, and this is most notable during the brief period in which Maria is living in London with Darnford following their escape from the madhouse.[70] One consequence of Maria's happiness at this point is a broadening of her social feelings, which are stimulated rather than excluded by her romantic fulfilment. Finding virtues in those she might previously have disregarded, Maria learns to acknowledge her fellow creatures through a budding of her real affections. This language of natural growth is central not just to Maria's experience of happiness but also to her meditations on it. A long paragraph contrasting real happiness with 'the reveries of a feverish imagination' turns on a contrast between 'roving through nature at large' and sporting in artificial 'gardens full of aromatic shrubs'.[71] Real happiness is not only like freely 'roving through nature', but also partakes of nature's organic qualities: its affections are 'buds pregnant with joy' which 'branch out with wild ease'. Artificial happiness, by contrast, offers an 'insipid uniformity which palls', its Rasselasian landscape fencing off sorrow and making us doze by a lake unruffled by the winds of change which will never be allowed to arrive.

This contrast between the dully constrained and the freely roaming also illuminates the difference between Maria's animated sensibility and Rousseau's genius. Unlike the man proclaimed by Darnford as the '"Prometheus" of sentiment', Maria's sensibility is not limited to acts of representation or fictional creation. When she chooses to write, she turns her energies not to a novel, such as might awaken 'feverish dreams of ideal wretchedness or felicity', but rather to a letter of improvement to her daughter, relating her own attempt to scale the walls of her confined valley, rove through nature and find fulfilment in social feeling. Rather than focusing her ability to animate on Darnford, in a repetition of Pygmalion's attempt to bring a statue to life, or in an operation of fevered fancy like the asylum inhabitants' excessive imaginative responses to every object, Maria's new sense of social connection allows her sensibility to animate her world, turning her experience of it into something like roving through nature, so that its joys can 'bud forth' with 'wild ease'. Proclaimed in the book's early stages as possessing the 'electric spark of genius', what has been animated for Maria by the end are the 'real affections of life', 'allowed to burst forth' with 'all the sweet emotions of the soul'. If the pruning, refining, cultivating language of conduct books finds its corollary in the artificial garden of false 'felicity', Maria's animating sensibility finally promises fulfilment and happiness in a wilder, less constrained, creation.

This account of sensibility, whilst produced through the temporary catalyst of romantic happiness, in fact rewrites the connection between imagination and love so ironically traced earlier in the text. Whereas a kind of false love can be fed by artificial fancy, 'real affections' neither produce nor depend on such feverish depictions, but rather on the imagination and sensibility manifested in social feeling: the recognition of others as fellow creatures identified as the primary act of sympathy in Smith's *Theory of Moral Sentiments*. The imagination recovered by the end of the text is not a feverish fancy which depicts ideal qualities abstracted from real existence, but an ability to understand and communicate with others ultimately stemming, in an inversion of Smith, not from their pain but from one's own feeling. Its writerly corollary is not the Rousseauean fiction but the maternal letter. Such a formulation of the crucial social context of sensibility's vital power erases the potential egotism and spurious self-sufficiency of Maria's moment of sublime self-realisation on leaving her husband, replacing it with a sense of proper social interdependence such as is exemplified again when (in one of the endings Wollstonecraft meditated) she is rescued from near-death by Jemima to resume her maternal duties with the support of a female companion.

The improvement of Maria's mind is thus finally revealed to be bound up with a new account of sensibility's animating powers, manifested not in fanciful romantic visions but in social and ultimately maternal affection. Writing her letter to her daughter, wishing for the 'improvement' which is 'ever present' as she writes, Maria asks her reader not to indulge in the kind of imaginative reading which she herself practises in the asylum, or indeed that Rousseau offers with his fictional portrayal of the sentiments. Rejecting the endless instructions and minutiae of the conduct book, she asks her daughter only for the 'energy of character' which will enable the 'buds' of real affection to flourish even in such a world as her mother has negotiated.[72] By the end of her own story, the insistence on the powers of animation which runs through the tale takes a terribly literal turn, as Maria's attempted suicide is pitched against Jemima's revivifying reminder of her maternal role. As so often in women's writing of this period, death threatens, but unlike Fidelia, Sibella and Mary, through Jemima's actions Maria is finally able to find a resource within herself capable of enabling her to live again. In animation's final appearance in this text, it is shorn of higher aspirations to sublime virtues and flowering genius, and is announced simply as what was denied to Wollstonecraft herself: the desire of a mother to 'live for my daughter'.

Maria's revival, in at least one of the endings sketched by Wollstonecraft to this unfinished fiction, with its dramatic transmutation of one possible fate awaiting 'woman' – death – into its opposite, is effected not through the powers and procedures of the medicalised resuscitation recommended by Cullen, Hunter and others, but through the animating force of maternal sensibility. As with the stories of Fidelia, Sibella, Mary and even Frankenstein's creature, a desire to imagine and write the tales of fulfilled lives lived freely, unthwarted by repression, confinement and restraint, finds its final expression in considerations of the cost, and possibility, of animation itself. Denied to Sibella and, in her final, eager anticipations of her end, to Mary; figured in religious terms for Fidelia, securing the possibility of living a life, even whilst in the very arms of death, is Maria's final achievement. Although writing a work of literary fiction, Wollstonecraft shares the concern of contemporary physiologists with understanding the possibility of life, animation and vitality, and she shares too their insistence on the importance of a sustaining, internal power, a 'principle of life', whose generative energies can even reverse the advances of death. Her characterisation of social sensibility in general, and maternal feeling in particular, as just such a power arguably completes the transmutation into literary discourse of the vitalists' century-long concern with the mysterious vital powers which define and sustain human life.

Conclusion: Eighteenth-Century Vitalism, Romanticism, Literature and the Disciplines

What does it mean for a study of eighteenth-century vitalism to conclude with a reading of Romantic-period writers? Where is the dividing line between an eighteenth-century vitalist language of nature, and fully fledged, transcendent Romantic organicism? And if, as the previous chapter asserts, a language of vitality and animation has a suggestive and fruitful figurative presence beyond natural philosophical contexts, what does this suggest about literature's relation to science at the end of the eighteenth century?

We saw in the previous chapter how a persistent language of vitality and animation in the literary writings of Romantic-period women writers is linked to their investigation of female oppression. The chapter concluded by suggesting that Wollstonecraft's preoccupation with genius and sensibility – so unresolved in her first novel, and so central to the second – develops into a language of vitality and animation through which Maria's personal fulfilment, if not emancipation, can finally be secured. Wollstonecraft's particular articulation of sensibility in this text, it was suggested, can be understood to inflect a contemporary fascination with a language of animation and vitality central to late eighteenth-century physiology and natural philosophy, but which – as the evidence from periodicals showed – existed too in the broader cultural realm. This language of vitality and animated life is articulated in part through references to organic nature, and an alternative reading of Wollstonecraft might indeed explore her concern with society, subjectivity and transcendence in relation to a traditional account of Romantic organicism. What difference does it make then, to read Wollstonecraft in the context of eighteenth-century vitalism? And what is the relationship of that context to a more recognisably Romantic one?

Such questions bring into review established scholarly and critical habits of periodisation in literary and historical study, as well as perceived relationships between literature, culture and the history of science. One purpose of this study has been to overturn an outdated sense of the eighteenth century as an age of mechanism, presided over by the twin gods of (to some eyes, unrecognisably mechanical accounts of) Newton and Locke.[1]

207

In this old critical story, an eighteenth century which is implicitly dry and barren is clearly distinguished from the more fruitful and welcome flowering of Romanticism, with the organicism which accompanies and defines it. But while challenging such outdated depictions, this study has asserted the distinct nature of a non-mechanical eighteenth century from Romantic organicism. The Introduction argued that vitality in an eighteenth-century context is both a theoretical supposition deployed in scientific discourses – to provide a non-mechanical, non-chemical answer to the question of the nature of life – and a suggestive language of vitally animated nature which informs numerous discourses beyond natural philosophy. In both cases, this is distinct from the more totalised transcendent organicism of the Romantic period where, as Charles Armstrong and Tilottama Rajan have recently reminded us, it operates as an 'all-pervasive' figure, so that organic images, analogies and structures inform Romantic depictions of things as various as the sacred book, the limit-experience (the sublime) and community.[2] But does an eighteenth-century vitalism give way to later organicism, or persist as part of it – and if there is a transition between them, does it neatly reflect our literary-historical distinction between the eighteenth century and Romanticism? How can we begin to separate these modes of thinking about nature, and how do we relate an account of them to our sense of the literary-historical period?

It is certainly clear that a renewed attention to the specificities of eighteenth-century natural philosophy can operate to dismantle a traditional, exclusive, association of animation, creativity and the organic with Romanticism. Critical accounts of the Romantic period have long emphasised animation, vitality and other organic tropes which appear in writings on (or manifestations of) poetic creativity, or in descriptions of imaginative powers. Equally, organic sensibility, creativity and genius are recognised as central terms in Romantic-era writing, and as such are usually discussed as exemplary instances of the larger aesthetic concerns of the period. But – and hence the problem or possibility of 'pre-Romanticism' – these are also well established as powerful terms in aesthetic discourse well before the late eighteenth century. Romanticism might be understood as representing a step-change from the writing and thinking of an earlier period – as in, for instance, the conventional distinction of the Romantic imagination from earlier versions by its ability not merely to combine received images (as in Addison or Locke's accounts of the faculty), but to create and animate anew – and clearly, acts of creating and animating exert a crucial fascination over writers of this era.[3] But where a more traditional literary criticism offered a formal history of such aesthetic trends and developments relatively divorced from historical or material context, the new critical historicism of recent decades considers aesthetic output and values alongside the political agendas, historical forces and social ideologies of their day.[4] The same historicising impulse has also produced important work on the relationship between contemporary science (itself understood not as an intellectual endeavour abstracted from the politics and pressures of its moment, but as a product and embodiment of it) and literary writing. The preoccupation of eighteenth-century natural philosophy with the forces which animate nature, and their particular focus through a vitalist science of life, thus offers a potential alternative context for reconsidering the favourite tropes of Romantic writing, and the origins of Romantic organicism. Further, the evidence offered by eighteenth-century natural philosophy and scientific thought, might usefully enable a fruitful review of period divisions which are still essentially founded on literary, cultural and historical change alone.

Considered narrowly, vitalism and organicism can be readily distinguished. In its strictest sense, vitalism is a theory which understands life in terms other than physical, chemical or mechanical; organicism, strictly defined, describes the belief that biological processes in plants and animals should be understood by reference to the dynamic mutual workings of parts and whole – not by reference to a mysterious additional force.[5] Organicism emphasises the organisation of structure and parts, and understands the operation of the whole in terms of unity of, and harmony between, its constituent elements – which is why plants, whose structures are more accurately described as a 'congeries of parts', presented such a problem for Coleridge's organicism (in contrast to Erasmus Darwin, for whom, as we saw in Chapter 5, plants were central to his exposition of the vitality of nature).[6] Thus whilst an eighteenth-century vitalist language of nature might appear to resemble a Romantic predilection for models drawn from organic nature, especially when such a language is deployed away from specific and precise scientific contexts, it is nevertheless grounded on a distinct understanding of the natural world.

Of course, continuities can be plotted between eighteenth-century vitalist theories and Romantic organicism. Romantic concern with organic form and organic unity is often understood through reference to the influence of German *Naturphilosophie* on English writers and thinkers, especially Coleridge, at the beginning of the nineteenth century, but an alternative, or supplementary, genealogy of influence might also be plotted: perhaps one which asserts the influence of John Hunter, whose ideas arguably contribute to Romantic paeans to 'one life', and who is praised at the beginning of Coleridge's unfinished *Theory of Life*, a text which, as Rajan argues, is a development of Hunter's attempts to identify a vital principle.[7] Indeed, the example of Coleridge, with his synthesising urges and wide-reaching intellectual interests, risks making an attempt to distinguish vitalism from organicism appear sterile and atomistic: precisely against the spirit in which he worked (Rajan has described his *Theory of Life* as an attempt to 'make friends' between the implications of Hunter's work and idealist German philosophy). What Armstrong has described as the 'romantic predilection for organic vitalism', its perception of an active universe animated by unknown, immaterial forces, can just as easily be traced back to an eighteenth-century English philosophical tradition as a nineteenth-century German one.[8] As Paul Hamilton has asserted, the associationism and pantheism which is connected so closely with Romanticism's organicism, stems from a tradition of what he terms 'spiritualised' English materialism, traceable, via Darwin and Priestley, back to Locke: a philosophical lineage closely intertwined with that of vitalism.[9] Equally, as we saw in Chapter 6, vitalism did not fade away as the Romantic period advanced, but maintained a controversial presence in English physiology well into the nineteenth century, reaching the height of its public awareness with the Abernethy–Lawrence controversy. Here again, drawing the line between an earlier vitalism and a later organicism is not a simple matter – especially as vitalism is compatible with versions of organic thought.

It is worth asking how far this attempt to differentiate, or connect, vitalism and organicism is helped or hindered by established categories of historical periodisation long used – though also questioned – by historians and literary critics. Even if a 'hard' sense of Romantic literature has, in recent years, been modified, under a disinclination to theorise a singular 'Romanticism', so that critics are more likely to refer more generally to 'writing of the Romantic period', the question of the Romantic is still somehow present, and is often linked to a persistent sense of the difference of this period from what preceded it. Alternative recent attempts to, variously, theorise the 'long eighteenth century' or the 'Romantic century' of 1750–1850 have not resolved this question, and the term 'pre-Romantic' has not gained real critical traction.[10] At the same time, recent interdisciplinary studies, such as the many recent explorations of the relationship between Romantic-era science and literature, have arguably only reinforced such established divisions in historical periods, which often remain in place too institutionally, in curricula, or in publishers' catalogues. Whilst interdisciplinary studies productively collapse the boundaries between literature and other forms of writing or cultural activity, its historicist methods can often, by contrast, depend on, rather than question, established divisions of historical period. But does evidence from the history of science support or trouble such distinctions in literary period?

Old physiology, new physiology: vitalism pre- and post-1800

Recent work by historian of science Andrew Cunningham throws fascinating light on these concerns. Reviewing the history of anatomy and physiology in and before the eighteenth century, Cunningham argues that in the years around 1800 new forms of these sciences emerged which were sufficiently different from earlier practices to constitute a radical disciplinary shift or change. Cunningham's argument dovetails with work by other historians of science, many of whom have pointed to the early years of the nineteenth century as the origin of modern scientific disciplines; Cunningham's work however is particularly compelling for this study not only in its detail, but also for the relevance of its attention to the history of physiology for our related concern with vitalism and historical period.

Cunningham describes how physiology prior to 1800 is a fundamentally different discipline from the experimental physiology which emerges in the early years of the nineteenth century. In his account, the central concern of what he calls 'old' physiology is to produce an essentially theoretical 'narration' of the causes of bodily motion, drawing on knowledge about the structure and parts of the body derived from anatomy. 'Old' physiology is thus a theoretical rather than an experimental discipline. 'New' physiology by contrast begins with experimentation and is explicitly a materialist science. In pre-1800 physiology, the story told about 'the ultimate source of motion and change in the body' is derived from whichever natural philosophical explanation is currently in favour and might be mechanical, animist or vitalist.[11] Vitalist physiology is thus described here as part of a theoretical narrative about the cause of bodily motion which builds on knowledge of bodily parts and structure derived from anatomy. Albrecht von Haller, a central figure in Cunningham's study, and a dominant figure in both eighteenth-century physiology and anatomy, thus described his physiology as 'animated anatomy', a phrase which Cunningham says might indeed describe the whole endeavour of pre-1800 physiology.[12]

Connected to this argument about 'old' physiology is one about 'old' anatomy which, like old physiology, also gave way to the new experimental physiology. Cunningham describes 'old' anatomy as distinct and distinguishable from physiology but often practised alongside it, even by the same person: he gives Thomas Willis and von Haller as two examples of individuals who practised both these disciplines whilst understanding them as separate endeavours. One sign of the significance of the shift from old anatomy to experimental physiology is indicated by the difference in the term 'organic' between the two disciplines. In 'old' anatomy, 'organic' meant 'belonging to the organ', and, in a field where bodily organs were understood to be animated by the soul, was strongly assimilated to such notions. In the new experimental physiology, by contrast, organic takes on a very different meaning associated with the notion of life (in contrast to 'inorganic'). A shift takes place in the understanding of the nature of life itself: as Cunningham says, '[a]n "organised body" in the new sense is alive because of the interrelationship of its constituent parts, and not because of its possession of discrete organs serving the soul'.[13] Organic in our modern sense – and organicism in the Romantic sense – are thus clearly post-1800 usages.

Cunningham's distinction between anatomy and physiology also throws light on the particularity of vitalism in eighteenth-century physiology. As he describes it, anatomy in this period is an experiment science, for which physiology provides the theory. The central theoretical question, as von

Haller's description of physiology as 'animated anatomy' implies, was how the static bodily structure moved or was 'animated'. This question produced assertions like this from Thomas Willis, who described the inner parts of even the smallest and 'vilest' animals as 'Hearths and Altars for the continuing Vital Fire' which animates the body.[14] Or, in a later example, when Galvani discovered electricity, he understood himself to be performing anatomical experiments which would make tangible the unknown principle of life about which physiologists had been speculating – and the phenomenon he discovered was indeed claimed by some to be the vital animating principle of life.[15] In this account of 'old', eighteenth-century physiology, vitalism is a theoretical speculation – a 'narration' in von Haller's words – about bodily animation; it is a search for the unknown principle which animates the known and visible parts and structure of the body. With the emergence post-1800 of experimental physiology, however, such theoretical speculations are no longer required: old physiology looks like a 'long and tiresome romance'.[16] To the extent that vitalism continued beyond 1800, it is no longer framed in this specific way, in relation to this specific disciplinary question about how a structure might be animated – even whilst questions about the nature of life and its presence and operation in the living organism remain.

Cunningham locates the rise of the new experimental physiology in France in the very early years of the nineteenth century, and states that 'old' physiology continued longer in England, in part due to political hostility to France. He also asserts that these new disciplinary forms are in various ways linked to the old, for instance in their deployment of old forms of language, or in their production of disciplinary histories (which Cunningham terms 'disciplinary cleansing') which obscure past disciplinary concerns by overwriting them with new ones.[17] Nevertheless, his is a powerful account of fundamental change in scientific discipline and practice, which surely must be understood as having consequences for vitalism and languages of animation post-1800, ones whose implications for literary Romanticism, and its difference from an earlier period, have not to date been fully considered. This must be especially true for Romantic writers like Percy Shelley and Coleridge, who had strong interests in the scientific developments of their day. At the same time, the continuity of language, as Cunningham describes, and the interests of a new discipline in cloaking itself in the dress of the old, means that such differences might be hard to pinpoint: especially when dealing not directly with the scientific texts themselves, but with literary and cultural responses to these ideas more broadly. At the level of a more general cultural reception, shifts in scientific paradigms, especially if some continuity of language and concept is present, might be less readily discernible. But that is not to suggest that they are not there.

What does all this mean for our question about eighteenth-century vitalism and its difference from Romantic organicism? On the one hand, it has been the effort of this book to establish the existence of vitalist thinking, and vitalist language of nature, within eighteenth-century natural philosophy, and to show how this informed other discourses. It has argued that a strong tradition of vitalism, within natural philosophy and physiology in particular, though by no means limited to it, and originating in responses to post-Newtonian matter theory, made its presence felt in other discourses, including literary and philosophical. On the other, disciplinary changes in the early nineteenth century suggest that Romantic organicism, whilst inevitably continuous in some ways with earlier thought, can also be understood as something new. Insofar as they are 'literary' writers usually contextualised within a periodisation drawn by cultural historians and critics, Cunningham's particular emphasis on the early nineteenth century as a scientific turning point carries consequences for writings by Wollstonecraft and others in the 1790s, as considered in the final chapters of this book. Throughout debates about periodisation, the pre-Romantic, the 'Romantic century' and so forth, the 1790s are almost always considered part of a 'Romantic' era, but this is a definition which, being concerned primarily with literary, cultural and perhaps political activity, pays little attention to accounts of the rise of the modern scientific disciplines dated by most commentators to the early nineteenth century.[18] However, such writing might be considered differently where – as in this book – a scientific context is relevant. Indeed, if considerations about the reordering of scientific disciplines are brought into play, and if science post-1800 is viewed as significantly different from that pre-1800, the possibility must be considered that the vitalist language of 1790s writing should be considered in a different category, and as perhaps significantly different, from Romantic writing post-1800. Considered in this way, the vitalist language of Darwin, Thelwall, Wollstonecraft and others might best be understood (as this book has attempted to do) by drawing a retrospective context which looks back to the persistent thread of vitalism in eighteenth-century natural philosophy – rather than attempting to assimilate them to a mature Romantic organicism which they at best only anticipate. A distinction between eighteenth-century Romantic-era writers and nineteenth-century ones, whilst sounding somewhat academic, might thus in fact open a significant new context for viewing their writing in a new, more historically – and, importantly, scientifically – accurate light.[19]

The discipline of literature

To conclude this work, I would like to view these questions from another perspective – one which brings the issue of disciplinarity into engagement with the history of literature itself. Here again the picture is complex, but a well-established story shows that the period around the beginning of the nineteenth century witnessed significant transformation not only in scientific, but also literary discourses. As Mary Poovey has recently shown, and as this study has evidenced, a relative lack of discursive specialisation existed in eighteenth-century Britain – one effect of which, as this study has argued, was the relatively easy cross-fertilisation of ideas, terminology, concepts and language between different areas of intellectual enterprise (between, say, vitalist physiology and political economy). Specifically, as Poovey asserts, there was a 'continuum that linked writing about the natural world to writing about human affairs', observable, for instance, and as Chapter 6 explored, in the periodicals, where 'writing about the natural world was published ... side by side with writing about literature, politics, ethics'.[20] As Poovey notes – an observation on which this study is predicated – this meant that 'it was easy for writers on all subjects to share a common language or to borrow terminology from each other'. This situation, however, did not hold, as in both areas with which we are concerned, literary and scientific, moves towards specialisation took place in the first decades of the nineteenth century. This manifested itself, for example, in the identification of specific, narrow, branches of scientific enquiry (such as experimental physiology, as discussed by Cunningham, or biology, in another well-known example). It also prompted the founding of specialised scientific societies and journals pertaining to the new scientific disciplines, and in the formulation of new, technical languages appropriate to such newly defined intellectual endeavours. Famously, the terms 'scientist', and 'science' in its specialised modern usage, were coined only from the mid-nineteenth century – precisely in an attempt to describe the *continuities* between different scientific endeavours which the fracturing just described threatened to eclipse. On its part, the term 'criticism' had been in use from the mid-eighteenth century, but debates over the identity, meaning and role of 'literature', 'criticism' and 'the critic' were ongoing, whilst the nature, place and social purpose of this kind of writing, and writing about writing, continued to be defined. Wordsworth's 'Preface' to the *Lyrical Ballads*, for instance, an obvious landmark in this process, was preoccupied with defining the language of poetry (which paradoxically would be that of 'common man'), as well as with distinguishing between the 'man of poetry' and the 'man of science', and adjudicating between their counterclaims for cultural and critical authority. In what Poovey describes as a 'complex cultural process', what we are familiar with today as 'literature' was over this period 'gradually separated from other kinds of writing', and invested with such notions as aesthetic autonomy, organic harmony, unity and so forth, and 'criticism' was distinguished as another, distinct and particular kind of writing about 'literature', with its own recognisable technical vocabularies and methods appropriate to it as a specific and recognisable discipline.[21]

This move, across the disciplinary spectrum, from the close relationship which existed between diverse Enlightenment practices of knowledge, to the formation of more specialised and distinct discourses, was manifested in numerous ways, including, at the institutional level, in university curricula and appointments; the period from the late eighteenth century onwards, for instance, saw the establishment of chairs in newly defined areas of study, such as political economy, or rhetoric and belles lettres, or, as it was later termed, literature. But with this new separation, the question of the relationship between, say, literature and the sciences, required redefinition, at a time when those disciplines themselves were undergoing the processes of self-definition. In the version of that redefinition and self-definition described by Poovey and others, the language of science – specifically the natural historical language of organicism – is co-opted by literature in a crucial moment of self-theorisation, in a move whose consequences remain with literary scholars today. As Poovey shows, Coleridge – a crucial figure in the self-formulation of literature at this time – borrows a language of organic form and organic unity to theorise the imagination and the artwork, and the relationship between literature, artist and reader, bequeathing literary criticism a 'founding' metaphor of organicism which has become so naturalised (a term which is itself an organicism) that it largely goes unnoticed.[22] Reinvigorated by Henry James and by mid-twentieth-century New Criticism, the organic language and model deployed in the study and analysis of literary texts has survived even the ebb and flow of later critical fashions and methodologies.

What does all this mean in terms of our particular question about vitalism and organicism? How does a Romantic and post-Romantic literary culture which uses organicism as (in Poovey's term) its 'model system' differ from an eighteenth century in which a language of vital nature is deployed across relatively undifferentiated knowledge practices? At the very least, Poovey's argument contributes another way to address the vexed question of periodisation in literary history, outlined earlier. That question might be answered not only by attempting to pinpoint a specific shift between modes of knowledge where observable continuities also exist, but also by identifying changes in the cultural roles, locations and usages of scientific and literary discourses at the turn of the nineteenth century. It suggests too that the nature of both literature and science themselves, and their relation to other forms of knowledge, need to be understood differently in the post-eighteenth-century era. Whilst, as ongoing and historically related practices, continuities clearly exist between organic sciences in the eighteenth century and after it, literary culture's relationship to the knowledge offered by them is markedly different – as Wordsworth's vexed preoccupation with the suddenly pressing question of the difference between the 'man of science' and the 'man of poetry' suggests. All of this means that the prevalence of organic language and models within literature, and criticism, in the Romantic and post-Romantic era should not blind us to the specificity and historical difference both of a vitalist language dominant in many discourses before that, and of the particular nature of its relationship to a range of other discourses, including what we now term literary writing. To make these observations is not to assert that Romantic organicism's difference from an earlier language of vital nature is simply a crude function of a shift in disciplinary structures of knowledge, but rather that that shift reorganised the relationship of science and literature (as well as what those practices were, and how they described themselves), so that an earlier, pre-existing relationship became less evident and visible just as literature underwent a moment of self-realisation. It has been the attempt of this book to uncover what has been hidden in that eclipse, and recover an earlier language, not of the organic, but of a vital nature, in writing of all kinds, before that moment of literature's self-incarnation.

Notes

Introduction

1. *Letters of Samuel Taylor Coleridge*, ed. E. L. Griggs (Oxford: Clarendon Press, 1956), vol. 1, pp. 294–5. For Coleridge and Thelwall, see Ian Wylie, *Young Coleridge and the Philosophers of Nature* (Oxford: Clarendon Press, 1989), pp. 123–6, and Trevor H. Levere, *Poetry Realized in Nature: Samuel Taylor Coleridge and Early Nineteenth-Century Science* (Cambridge University Press, 1981), p. 208. Coleridge refers to the 'mechanical solutions' of the eighteenth century in a discussion of the physiologist John Hunter, in his *Theory of Life*. See *The Collected Works of Samuel Taylor Coleridge*, vol. 11, *Shorter Works and Fragments*, ed. H. J. Jackson and J. R. de J. Jackson (London: Routledge, 1995), p. 486.
2. The phrase 'language of nature' is used by Peter Hanns Reill in his *Vitalizing Nature in the Enlightenment* (Berkeley: University of California Press, 2005); see also Ludmilla Jordanova, ed., *Languages of Nature: Critical Essays on Science and Literature* (London: Free Association, 1986). The term 'science', which did not gain its current meaning until the nineteenth century, is used anachronistically throughout this book to refer to the wide and various range of natural philosophical and experimental researches and activities which we now understand as scientific in the modern sense.
3. Alexander Pope, 'An Essay on Man', in *The Poems of Alexander Pope*, ed. John Butt (1963, repr. London: Routledge, 1989), III, 13–26 (p. 526).
4. For Pope's strategic contradiction, see David Fairer, 'Introduction', in *Pope: New Contexts*, ed. D. Fairer (Hemel Hempstead: Harvester Wheatsheaf, 1990), p. 6. For a full account of the interpretative difficulties surrounding the *Essay on Man*, as well as a reading of the text as deliberately resisting 'polite' communication, see Stephen Copley and David Fairer, '*An Essay on Man* and the Polite Reader', in *Pope: New Contexts*, ed. Fairer, pp. 205–24; and for Coleridge's failure to appreciate such textual subtleties, see John Whale, 'Romantic Attacks: Pope and the Spirit of Language', in *Pope: New Contexts*, ed. Fairer, pp. 153–68 (p. 160).
5. For these and other examples, see the *OED*. For *The Ladies Dispensatory*, see Vivien Jones, ed., *Women in the Eighteenth Century* (London: Routledge, 1990), p. 83.
6. 'Essay on Man', III, 115–18 (p. 529).
7. On Newtonian and post-Newtonian matter theory, see P. M. Heimann, '"Nature is a Perpetual Worker": Newton's Aether and Eighteenth-Century Natural Philosophy', *Ambix*, 20 (1973), 1–25; Heimann, 'Voluntarism and Immanence: Conceptions of Nature in Eighteenth-Century Thought', *Journal of the History of Ideas*, 39 (1978), 271–83; Heimann and J. E. McGuire, 'Newtonian Forces and Lockean Powers: Concepts of Matter in Eighteenth-Century Thought', *Historical Studies in the Physical Sciences*, 3 (1971), 233–306; and Alan Gabbey, 'Newton, Active Powers, and the Mechanical Philosophy', in *The Cambridge Companion to Newton*, ed. I. Bernard Cohen and George E. Smith (Cambridge University Press, 2002), pp. 329–57. For Newton's impact on physiology, see Anita Guerrini, 'James Keill, George Cheyne, and Newtonian Physiology, 1690–1740', *Journal of the History of Biology*, 18 (1985), 247–66, and Shirley Roe, 'The Life Sciences', in

The Cambridge History of Science, vol. 4, *Eighteenth-Century Science*, ed. Roy Porter (Cambridge University Press, 2003), pp. 397–416.

8. See John P. Wright, *The Sceptical Realism of David Hume* (Manchester University Press, 1983), p. 162.

9. On post-Lockean debates, see John Yolton, *Thinking Matter: Materialism in Eighteenth-Century Britain* (Oxford: Blackwell, 1983).

10. See Sergio Moravia, 'From "Homme Machine" to "Homme Sensible": Changing Eighteenth-Century Models of Man's Image', *Journal of the History of Ideas*, 39 (1978), 45–60; François Duchesneau, 'Vitalism in Late Eighteenth-Century Physiology: The Cases of Barthez, Blumenbach and John Hunter', in *William Hunter and the Eighteenth-Century Medical World*, ed. W. F. Bynum and Roy Porter (Cambridge University Press, 1985) and Elizabeth L. Haigh, 'Vitalism, the Soul, and Sensibility: The Physiology of Theophile Bordeu', *Journal of the History of Medicine and Allied Sciences*, 31 (1976), 30–41. For vitalism in the European Enlightenment more generally, see Reill, *Vitalizing Nature*.

11. See Guerrini, 'James Keill, George Cheyne, and Newtonian Physiology', for claims of Cheyne's incipient vitalism in a slightly earlier period.

12. For the history of the Edinburgh Medical School, see J. B. Morrell, 'The University of Edinburgh in the Late Eighteenth Century', *Isis*, 62 (1971), 158–71; John R. R. Christie, 'The Origins and Development of the Scottish Scientific Community, 1680–1760', *History of Science*, 12 (1974), 122–41; Christopher Lawrence, 'The Nervous System and Society in the Scottish Enlightenment', in *Natural Order: Historical Studies of Scientific Culture*, ed. Barry Barnes and Steven Shapin (London: Sage, 1979), pp. 19–40; and A. L. Donovan, *Philosophical Chemistry in the Scottish Enlightenment: The Doctrines and Discoveries of William Cullen and Joseph Black* (Edinburgh University Press, 1975), ch. 3.

13. Innumerable physicians in this period had links to Edinburgh. These include Erasmus Darwin, Thomas Beddoes and John Brown, whose 'Brunonian' medicine, based on the theory of bodily excitability, was widely adopted in Germany and used as a basis for Romantic medicine. See Christopher Lawrence, 'The Power and the Glory: Humphry Davy and Romanticism', in *Romanticism and the Sciences*, ed. Andrew Cunningham and Nicholas Jardine (Cambridge University Press, 1990), pp. 213–27 (p. 215). William Hunter, brother of the eminent physiologist John Hunter, and a leading anatomist and physician in his own right, was also an early pupil and lifelong correspondent of Cullen.

14. Sharon Ruston, *Shelley and Vitality* (Basingstoke: Palgrave, 2005); Nicholas Roe, *The Politics of Nature: William Wordsworth and Some Contemporaries* (Basingstoke: Palgrave, 2002). For Romanticism and vitalism, see also Denise Gigante, *Life: Organic Form and Romanticism* (New Haven and London: Yale University Press, 2009), Richard Holmes, *The Age of Wonder* (London: Harper, 2008), and Clayton Koelb, *The Revivifying Word: Literature, Philosophy, and the Theory of Life in Europe's Romantic Age* (Rochester, NY: Camden House, 2008).

15. Alan Richardson, *British Romanticism and the Science of the Mind* (Cambridge University Press, 2001). On Romantic science, see, *inter alia*: Andrew Cunningham and Nicholas Jardine, eds, *Romanticism and the Sciences* (Cambridge University Press, 1990); Hermione de Almeida, *Romantic Medicine and John Keats* (Oxford University Press, 1991); Adrian Desmond, *The Politics of Evolution: Morphology, Medicine and Reform in Radical London* (Chicago University Press, 1989); Levere, *Poetry Realized in Nature*; Nicholas Roe, ed., *Samuel Taylor Coleridge and the Sciences of Life* (Oxford University Press, 2001); and Wylie, *Young Coleridge*.

16. Samuel Johnson, *The Rambler*, no. 168 (26 October 1751), 377–9 (p. 378). On animation in eighteenth-century poetry and criticism, see my 'The Science and Poetry of Animation: Personification, Analogy, and Erasmus Darwin's *Loves of the Plants*', *Romanticism*, 10:2 (2004), 191–208 (pp. 195–6).
17. Ruston, *Shelley and Vitality*, pp. 24, 15.
18. Ruston, *Shelley and Vitality*, p. 23. Haller introduced the terms 'sensibility' and 'irritability' to mid-eighteenth-century anatomy: see Andrew Cunningham, 'The Pen and the Sword: Recovering the Disciplinary Identity of Physiology and Anatomy before 1800. II: Old Anatomy – the Sword', *Studies in History and Philosophy of Biological and Biomedical Sciences*, 34 (2003), 51–76 (p. 66). Central to the debate between Whytt and Haller was the former's contention that the irritability of living tissue derived from a living principle within it. Haller preferred to explain muscular action by what Shirley Roe has described as 'animal mechanics'. Roe, 'Life Sciences', p. 403.
19. For Coleridge on Davy, see Nicholas Roe, 'Introduction', in *Samuel Taylor Coleridge*, ed. Roe, p. 12. For Coleridge's resistance to the modern term 'scientist', see Ruston, *Shelley and Vitality*, p. 186. Wordsworth's 'Preface' to the *Lyrical Ballads* also expounds on the relation of the 'Man of Science' and the 'Man of Poetry', in part as a response to the claims of Davy's Royal Institution lectures. See my 'The Science and Poetry of Animation'.
20. On the dangers of confusing materialism, mechanism and vitalism, see Richardson, *British Romanticism*, pp. 29–30, 69 and 192, n. 58.
21. Roy Porter and George Rousseau, 'Introduction: Toward a Natural History of Mind and Body', in *The Languages of Psyche: Mind and Body in Enlightenment Thought*, ed. G. S. Rousseau (Berkeley: University of California Press, 1990), p. 31. Peter de Bolla similarly sees this period as one 'of an emerging new understanding of the construction of the subject' in his 'Introduction' to *The Sublime: A Reader in British Eighteenth-Century Aesthetic Theory* (Cambridge University Press, 1996), p. 1. Dror Wahrmann's *The Making of the Modern Self: Identity and Culture in Eighteenth-Century England* (New Haven: Yale University Press, 2004) is the most recent examination of the eighteenth-century 'self'; see also Michael McKeon, 'Biography, Fiction, and the Emergence of "Identity" in Eighteenth-Century Britain', in *Writing Lives: Biography and Texuality, Identity and Representation in Early Modern England*, ed. Kevin Sharpe and Steven N. Zwicker (Oxford University Press, 2008), pp. 339–55; Felicity Nussbaum, *The Autobiographical Subject: Gender and Ideology in Eighteenth-Century England* (Baltimore: Johns Hopkins University Press, 1989); and Patricia M. Spacks, *Imagining a Self: Autobiography and Novel in Eighteenth-Century England* (Cambridge, MA: Harvard University Press, 1976).
22. Dennis Todd, *Imagining Monsters: Miscreations of the Self in Eighteenth-Century England* (University of Chicago Press, 1995).
23. On the *Essay* as 'natural history', see James G. Buickerood, 'The Natural History of the Understanding: Locke and the Rise of Facultative Logic in the Eighteenth Century', *History and Philosophy of Logic*, 6 (1985), 157–90. On Locke's discussion of personal identity, see Christopher Fox, *Locke and the Scriblerians: Identity and Consciousness in Early Eighteenth-Century Britain* (Berkeley: University of California Press, 1988).
24. John Locke, *An Essay Concerning Human Understanding*, ed. Peter H. Nidditch (Oxford: Clarendon Press, 1975), p. 341.
25. Locke, *Essay*, p. 341.
26. Locke, *Essay*, p. 330. On Shelley's discussion of the oak tree, see Ruston, *Shelley and Vitality*, p. 8.

27. Locke, *Essay*, pp. 330–1.
28. Locke, *Essay*, p. 331.
29. Locke, *Essay*, pp. 331–2.
30. Joseph Butler, 'Of Personal Identity', in *The Works of Joseph Butler*, ed. W. E. Gladstone, 2 vols (Oxford: Clarendon Press, 1896), vol. 1, p. 392. Quoted in Fox, *Locke and the Scriblerians*, p. 10.
31. On Willis, see Andrew Cunningham, 'Old Anatomy', pp. 64–5.
32. David Hume, *An Enquiry Concerning Human Understanding*, ed. L. A. Selby-Bigge, rev. P. H. Nidditch (Oxford: Clarendon Press, 1975), p. 66.
33. Isaac Newton, *Mathematical Principles*, trans. Andrew Motte, rev. Florian Cajori (Berkeley: University of California Press, 1934), p. 547.
34. Isaac Newton, *Opticks, or A Treatise of the Reflections, Refractions, Inflections, and Colours of Light* (London, 1730, repr. London: Bell and Sons, 1931), p. 401.
35. My argument here and in the next paragraph runs parallel with that in my 'Feigning Fictions: Imagination, Hypothesis, and Philosophical Writing in the Scottish Enlightenment', *Eighteenth-Century: Theory and Interpretation*, 44:2 (2007), 149–71. On vitalism as a response to Hume's sceptical attack on the methods of natural philosophy, see Reill, *Vitalizing Nature in the Enlightenment*.
36. Robert Whytt, 'Essay on the Vital and Other Involuntary Motions of Animals', in *The Works of Robert Whytt* (Edinburgh, 1768), p. v.
37. Whytt, 'Essay', pp. v–vi.
38. William Cullen, *A Letter to Lord Cathcart, President of the Board of Police in Scotland, Concerning the Recovery of Persons Drowned and Seemingly Dead* (Edinburgh, 1776), pp. 25–6. For interest in resuscitation and reanimation across Europe in the late eighteenth century, see Reill, *Vitalizing Nature*, pp. 171–82, and Sean M. Quinlan, 'Apparent Death in Eighteenth-Century France and England', *French History*, 9 (1995), 27–47.
39. Cullen, *Letter*, pp. 2–3.
40. Locke, whom he encouraged his medical students to read, was a valued authority for Cullen. See Donovan, *Philosophical Chemistry*, p. 42.
41. *The Spectator*, 9 August 1714, no. 578, in *The Spectator*, ed. Gregory Smith, 4 vols (London: Dent, 1963), vol. 4, pp. 315–19.

1 Forms of Enlightenment: Embodied Beings in Eighteenth-Century Scotland

1. John Arbuthnot, *Gnothi Seauton: Know Yourself* (London, 1734), repr. in *Literature and Science 1660–1834*, 8 vols (London: Pickering and Chatto, 2003), vol. 2, ed. Clark Lawlor and Akihito Suzuki, pp. 79–88.
2. Christopher Fox discusses Scriblerian satire and its philosophical targets in *Locke and the Scriblerians*.
3. Arbuthnot, *Know Yourself*, ll.1–6.
4. On this, see Yolton, *Thinking Matter*, and Todd, *Imagining Monsters*, pp. 118–25.
5. John Arbuthnot, *An Essay Concerning the Nature of Aliments* (London, 1731), p. v.
6. Arbuthnot, *Essay*, pp. 18, 63, 167–8.
7. Arbuthnot, *Know Yourself*, ll. 24–8.
8. John Armstrong, *The Art of Preserving Health* (London, 1744), repr. in *Literature and Science 1660–1834*, vol. 2, ed. Lawlor and Suzuki, Book 4, ll. 12–13, 16–22.

9. An extensive literature exists on the transformation of moral philosophy in Scottish universities at the beginning of the eighteenth century, and the influence of natural philosophy on the Scottish Enlightenment more generally. See Michael Barfoot, 'Hume and the Culture of Science in the Early Eighteenth Century', in *Studies in the Philosophy of the Scottish Enlightenment*, ed. M. A. Stewart (Oxford: Clarendon Press, 1990), pp. 151–90; Ronald G. Cant, 'Origins of the Enlightenment in Scotland: The Universities', in *The Origins and Nature of the Scottish Enlightenment*, ed. R. H. Campbell and Andrew S. Skinner (Edinburgh: John Donald, 1982), pp. 42–64; Roger L. Emerson, 'Natural Philosophy and the Problem of the Scottish Enlightenment', *Studies on Voltaire and the Eighteenth Century*, 242 (1986), 243–91, and Emerson, 'Science and Moral Philosophy in the Scottish Enlightenment', in *Studies in the Philosophy of the Scottish Enlightenment*, ed. Stewart, pp. 11–36; Peter Jones, ed., *Philosophy and Science in the Scottish Enlightenment* (Edinburgh: John Donald, 1988); Christine M. Shepherd, 'Newtonianism in Scottish Universities in the Seventeenth Century', in *Origins and Nature of Scottish Enlightenment*, ed. Campbell and Skinner, pp. 65–85; Paul Wood, 'Science and the Aberdeen Enlightenment', in *Philosophy and Science in the Scottish Enlightenment*, ed. Jones, pp. 39–66, and Wood, 'Science and the Pursuit of Virtue in the Aberdeen Enlightenment', in *Origins and Nature of the Scottish Enlightenment*, ed. Campbell and Skinner, pp. 127–49.

10. David Shuttleton argues that Arbuthnot was intimate with Scottish Tory Newtonians, and may have defended the Newtonian Archibald Pitcairne in Edinburgh disputes over the new science. See his '"A Modest Examination": John Arbuthnot and the Scottish Newtonians', *British Journal for Eighteenth-Century Studies*, 18:1 (1995), 47–62. See also Anita Guerrini, 'The Tory Newtonians: Gregory, Pitcairne, and their Circle', *Journal of British Studies*, 25 (1986), 288–311.

11. William Leechman, 'Preface: Giving some Account of the Life, Writings, & Character of the Author', in *A System of Moral Philosophy*, by Francis Hutcheson (Glasgow, 1755), pp. i–xlviii. On Hutcheson, see T. D. Campbell, 'Francis Hutcheson: "Father" of the Scottish Enlightenment', in *Origins and Nature of the Scottish Enlightenment*, ed. Campbell and Skinner, pp. 167–85.

12. Leechman, 'Preface', p. xiv.

13. David Fordyce, *Elements of Moral Philosophy* (London, 1754), pp. 19–20.

14. Fordyce, *Elements*, p. 64.

15. Francis Hutcheson, *A Short Introduction to Moral Philosophy, in Three Books* (Glasgow, 1747), p. 1; Leechman, 'Preface', p. xv.

16. Fordyce, *Elements*, p. 75.

17. Hutcheson, *Short Introduction*, p. 2.

18. David Hume, 'Abstract' to *A Treatise of Human Nature*, ed. L. A. Selby-Bigge (Oxford: Clarendon Press, 1978), p. 646; Hume, *The Letters of David Hume*, ed. J. Y. T. Greig, 2 vols (Oxford: Clarendon Press, 1932), vol. 1, pp. 32–3.

19. Hume, *An Enquiry Concerning Human Understanding*, p. 66.

20. John Sekora's *Luxury: The Concept in Western Thought* (Baltimore: Johns Hopkins University Press, 1977) remains a classic account of such anxieties.

21. My argument in this section has already appeared in similar form in 'Disability and Sympathetic Sociability in Enlightenment Scotland: The Case of Thomas Blacklock', *British Journal for Eighteenth-Century Studies*, 30:3 (2007), 423–38.

22. Joseph Spence, 'An Account of the Life, Character, and Poems of the Author', in *Poems by Mr. Thomas Blacklock* (London, 1756), p. xlv.

23. Katie Trumpener, *Bardic Nationalism: The Romantic Novel and the British Empire* (Princeton University Press, 1997), pp. 96–100; see also Edward Larrissy, 'The Celtic Bard of Romanticism: Blindness and Second Sight', *Romanticism*, 5:1 (1999), 43–57.
24. Spence, 'Account', pp. xlv–xlvi.
25. Hume, *Letters*, ed. Greig, I, pp. 200–1.
26. [Thomas Blacklock], entry on 'Blind', *Encyclopædia Britannica*, vol. 2, 2nd edn (Edinburgh, 1778), pp. 1188–204 (p. 1189).
27. [Blacklock], 'Blind', p. 1196.
28. Thomas Blacklock, *Poems on Several Occasions* (Glasgow, 1746), p. 26. There are minor changes to this poem in Blacklock's later collections of 1754 and 1793.
29. Blacklock, *Poems on Several Occasions* (Edinburgh, 1754), pp. 141–55 (pp. 144–5). This poem is not included in the 1746 collection.
30. For a discussion of the eighteenth-century construction of disability as difference, see Helen Deutsch and Felicity Nussbaum, eds, *Defects: Engendering the Modern Body* (Ann Arbor: University of Michigan Press, 2000), Introduction, and Nussbaum, *The Limits of the Human: Fictions of Anomaly, Race and Gender in the Long Eighteenth Century* (Cambridge University Press, 2003).

2 Generating Sympathy: Sensibility, Animation and Vitality in Adam Smith and Mary Wollstonecraft

1. Mary Wollstonecraft, *A Vindication of the Rights of Woman*, in *The Works of Mary Wollstonecraft*, ed. Janet Todd and Marilyn Butler, 7 vols (London: Pickering and Chatto, 1989), vol. 5, pp. 112–13.
2. Important secondary literature on physiology and sensibility includes 'Nerves, Spirits and Fibres: Towards the Origins of Sensibility', in George Rousseau, *Nervous Acts* (New York: Palgrave, 2004), Ann Jessie Van Sant, *Eighteenth-Century Sensibility and the Novel* (Cambridge University Press, 1993), John Mullan, *Sentiment and Sociability* (Oxford: Clarendon Press, 1990) and G. J. Barker-Benfield, *The Culture of Sensibility* (University of Chicago Press, 1992).
3. *Mary, A Fiction*, in *The Works of Mary Wollstonecraft*, ed. Todd and Butler, vol. 1, p. 7. Lady Kingsborough, Wollstonecraft's employer whilst she worked as a governess in Ireland, is frequently identified as the source for these women. For Claudia Johnson's observation that Eliza is both 'hypercorporeal ... unanimated by higher faculties of mind or spirit' and 'hypocorporeal', see her *Equivocal Beings: Politics, Gender and Sentimentality in the 1790s* (University of Chicago Press, 1995), p. 51.
4. For discussion along these lines, see Chris Jones, *Radical Sensibility* (London: Routledge, 1993), John Whale, 'Death in the Face of Nature: Self, Society and Body in Wollstonecraft's *Letters Written in Sweden, Norway and Denmark*', *Romanticism*, 1:2 (1995), 177–92, and Barbara Taylor, *Mary Wollstonecraft and the Feminist Imagination* (Cambridge University Press, 2003).
5. Wollstonecraft, *Vindication of the Rights of Woman*, p. 115.
6. Wollstonecraft, *Vindication of the Rights of Woman*, pp. 75–6. Godwin recorded the very real effortfulness of Wollstonecraft's composition of the *Vindication*, including her exhaustion at mid-point, in his *Memoirs of the Author of 'A Rights of Woman'*.
7. Wollstonecraft, *Vindication of the Rights of Woman*, p. 130.
8. Whale, 'Death in the Face of Nature', p. 177.

9. Mary Wollstonecraft, *A Vindication of the Rights of Men*, in *The Works of Mary Wollstonecraft*, ed. Todd and Butler, vol. 5, p. 31.
10. Wollstonecraft, *Mary*, Advertisement, p. 5.
11. On seventeenth-century moral philosophy, see Susan James, *Passion and Action: The Emotions in Seventeenth-Century Philosophy* (Oxford: Clarendon Press, 1997).
12. Adam Smith, *The Theory of Moral Sentiments*, ed. D. D. Raphael and A. L. MacFie (Oxford University Press, 1976), p. 158. Future references will be abbreviated to *TMS*.
13. Archibald Campbell, *An Enquiry into the Original of Moral Virtue* (London, 1734), pp. 29–30. Italics as in the original.
14. Fordyce, *Elements*, pp. 170–1.
15. Fordyce, *Elements*, p. 171.
16. *TMS*, p. 9.
17. On medicalised sensibility, see especially Mullan, *Sentiment and Sociability*.
18. For the differences between Hume and Smith on sympathy, see David R. Raynor, 'Hume's Abstract of Adam Smith's *Theory of Moral Sentiments*', *Journal of the History of Philosophy*, 22 (1984), 51–79.
19. David Hume, *Enquiries Concerning Human Understanding and the Principles of Morals*, ed. L. A. Selby-Bigge and P. H. Nidditch (Oxford: Clarendon Press, 1975), pp. 66 and 219–20.
20. *TMS*, pp. 13, 71.
21. *TMS*, p. 10.
22. Such 'spectral' effects of the sympathetic imagination may explain the fertile connections between sentimental and gothic literature.
23. Peter de Bolla, 'The Visibility of Visuality', in *Vision in Context: Historical and Contemporary Perspectives on Sight*, ed. Teresa Brennan and Martin Jay (London: Routledge, 1996), pp. 65–79.
24. Adam Smith, *Lectures on Rhetoric and Belles Lettres*, ed. J. C. Bryce (Oxford University Press, 1983), p. 58.
25. On theatrical metaphors in *TMS*, see David Marshall, *The Figure of Theatre: Shaftesbury, Defoe, Adam Smith and George Eliot* (New York: Columbia University Press, 1986).
26. *TMS*, pp. 10, 12, 30.
27. Adam Smith, 'The History of Ancient Physics', in *Essays on Philosophical Subjects*, ed. W. P. D. Wightman and J. C. Bryce (Oxford: Clarendon Press, 1980), p. 107.
28. Smith, 'History of Astronomy', in *Essays on Philosophical Subjects*, p. 41.
29. Smith, 'History of Ancient Physics', p. 116.
30. Smith, 'History of Astronomy', p. 105.
31. *TMS*, p. 87.
32. *TMS*, p. 339.
33. *TMS*, p. 329.
34. *TMS*, p. 83.
35. *TMS*, pp. 134–6.
36. *TMS*, p. 137.
37. *The Spectator*, no. 413, 24 June 1712.
38. *TMS*, pp. 110–11.
39. Smith, 'On the External Senses', in *Essays on Philosophical Subjects*, p. 159.
40. Smith, 'External Senses', pp. 161, 160.
41. Mary Shelley, *Frankenstein, or the Modern Prometheus*, ed. Nora Crook (London: Pickering and Chatto, 1996), p. 81.

42. John Barrell discusses the class associations of viewing landscapes in 'The Public Prospect and the Private View: The Politics of Taste in Eighteenth-Century Britain', in *Reading Landscape: Country-City-Capital*, ed. Simon Pugh (Manchester University Press, 1990), pp. 19–40.
43. Mary Wollstonecraft, *Letters Written During a Short Residence in Sweden, Norway, and Denmark*, in *The Works of Mary Wollstonecraft*, ed. Janet Todd and Marilyn Butler, 7 vols (London: Pickering and Chatto, 1989), vol. 6, p. 243.
44. Wollstonecraft, *Short Residence*, p. 245.
45. Wollstonecraft, *Short Residence*, p. 248.
46. Wollstonecraft, *Short Residence*, pp. 248–9.
47. Wollstonecraft, *Short Residence*, p. 246.
48. Wollstonecraft, *Short Residence*, p. 279.
49. Wollstonecraft, *Short Residence*, p. 262.

3 Labouring Bodies in Political Economy: Vitalist Physiology and the Body Politic

1. Adam Smith, *An Inquiry into the Nature and Causes of the Wealth of Nations* (Oxford University Press, 1976), p. 10. Subsequent references will be abbreviated to *WN*.
2. *WN*, p. 22.
3. *WN*, p. 23.
4. *WN*, pp. 47, 117.
5. *WN*, p. 100. Ramazzini's *Treatise* was published in English in London in 1705.
6. *WN*, pp. 117, 126.
7. *WN*, pp. 98, 93.
8. *WN*, p. 21.
9. 'The Principles which Lead and Direct Philosophical Enquiries: Illustrated by the History of Astronomy', in Adam Smith, *Essays on Philosophical Subjects*, ed. W. P. D. Wightman and J. C. Bryce (Oxford University Press, 1980), pp. 33–105.
10. For Mary Poovey, political economy's production of 'knowledge' as a combination of the specific and particular with the general and abstract makes it characteristic of what she identifies as 'the modern fact'. See her *A History of the Modern Fact: Problems of Knowledge in the Sciences of Wealth and Society* (University of Chicago Press, 1998).
11. See J. G. A. Pocock, *Virtue, Commerce, and History: Essays on Political Thought and History, Chiefly in the Eighteenth Century* (Cambridge University Press, 1985), esp. pp. 113–15; Albert O. Hirschman, *The Passions and the Interests: Political Arguments for Capitalism before its Triumph* (Princeton University Press, 1977).
12. Tom Furniss, *Edmund Burke's Aesthetic Ideology: Language, Gender and Political Economy in Revolution* (Cambridge University Press, 1993), p. 52.
13. Locke, *Essay*, p. 253.
14. Furniss, *Edmund Burke's Aesthetic Ideology*, p. 53.
15. Bernard Mandeville, *The Fable of the Bees: Or, Private Vices, Publick Benefits*, ed. F. B. Kaye, 2 vols (Oxford: Clarendon Press, 1924), vol. 1, p. 333.
16. My argument on Hume here has also appeared in my journal article, 'Feigning Fictions'.
17. Hume, *Treatise*, pp. 422–3.
18. David Hume, *Essays: Moral, Political and Literary*, ed. Eugene Miller (Indianapolis: Liberty Fund, 1985), pp. 269–70.

19. Hume, *Treatise*, p. 424.
20. Hume, *Essays*, p. 270.
21. Hume, *Essays*, p. 271. Smith's early *Lectures on Rhetoric and Belles Lettres* echo Hume's analysis here by similarly suggesting that, in modern commercial society, work and pleasure are mixed, in contrast with the sharp division between work and pleasure in early societies. However, the *Wealth of Nations* offers a different account of work, with its recognition on the tedium and mental atrophy brought about by repetitive, unvaried work. See *WN*, pp. 781–2.
22. Christine Battersby, 'Hume, Newton and the "Hill Called Difficulty"', in *Philosophers of the Enlightenment*, ed. S. C. Brown (Brighton: Harvester Press, 1979), pp. 31–55. On Hume's Newtonianism, see Nicholas Capaldi, *David Hume: The Newtonian Philosopher* (Boston: Twayne, 1975), and William A. Wallace, *Causality and Scientific Explanation*, 2 vols (Ann Arbor: University of Michigan Press, 1974), vol. 1, pp. 40–4. For Hume's critique of some foundational concepts of Newtonian natural philosophy, see Heimann and McGuire, 'Newtonian Forces', p. 263.
23. Hume, 'Of the Passions', in *Four Dissertations* (London, 1757, repr. Bristol: Thoemmes Press, 1995), p. 181.
24. See Barfoot, 'Hume and the Culture of Science'.
25. *WN*, pp. 99–100.
26. *WN*, p. 100.
27. *WN*, pp. 341, 99.
28. *WN*, p. 341. See also pp. 540 and 674.
29. Locke, *Essay*, p. 280; Furniss, *Edmund Burke's Aesthetic Ideology*, p. 54.
30. Hume, *Treatise*, p. 215.
31. Hume, *Treatise*, pp. 265, 267.
32. Hume, *Letters*, ed. Greig, vol. I, Letter 3, March or April 1734, pp. 12–18. The letter may never have been sent, and its intended recipient is the subject of some conjecture. Greig follows John Hill Burton, author of *The Life and Correspondence of David Hume*, 2 vols (Edinburgh, 1846), in seeing it as addressed to George Cheyne; however, Ernest Mossner suggests that the addressee was John Arbuthnot. See Ernest Campbell Mossner, 'Hume's Epistle to Dr. Arbuthnot, 1734: The Biographical Significance', *Huntington Library Quarterly*, 7 (1944), 135–52.
33. Hume, *Treatise*, pp. 268–9, 264.
34. Furniss, *Edmund Burke's Aesthetic Ideology*, p. 43.
35. My argument in this section has already appeared in similar form in my article, 'The Physiology of Political Economy: Vitalism and Adam Smith's *Wealth of Nations*', *Journal of the History of Ideas*, 63:3 (2002), 465–81.
36. S. Todd Lowry, 'The Archaeology of the Circulation Concept in Economic Theory', *Journal of the History of Ideas*, 35 (1974), 429–44; Andrea Finkelstein, *Harmony and the Balance: An Intellectual History of Seventeenth-Century English Economic Thought* (Ann Arbor: University of Michigan Press, 2000).
37. Daniel Defoe, *The Review*, VIII (16), 1 May 1711.
38. *WN*, p. 108.
39. *WN*, pp. 604–6.
40. *WN*, pp. 466–7.
41. *WN*, p. 472.
42. *WN*, p. 496.
43. *WN*, p. 343.
44. *WN*, pp. 673–4, 343.

45. Adam Smith, *The Correspondence of Adam Smith*, ed. Ernest Campbell Mossner and Ian Simpson Ross (Oxford: Clarendon, 1987), no. 161, p. 201.
46. Smith, *Correspondence*, no. 51, p. 69.
47. For Cullen's important medical practice, see Guenter B. Risse, 'Doctor William Cullen, Physician, Edinburgh: A Consultative Practice in the Eighteenth Century', *Bulletin of the History of Medicine*, 48 (1974), 338–51. On the role of Cullen's 'vitalist' account of nature as self-healing in his teaching, see Rosalie Stott, 'Health and Virtue: Or, How to Keep out of Harm's Way: Lectures on Pathology and Therapeutics by William Cullen c. 1770', *Medical History*, 31 (1987), 123–42.
48. On Bordeu and his link with the *Encyclopédie*, see Haigh, 'Vitalism, the Soul, and Sensibility'; see also Moravia, 'Homme Machine', and Reill, *Vitalizing Nature*, pp. 131–4.
49. On mechanist natural philosophy and the shift to a vitalist physiology, see Robert E. Schofield, *Mechanism and Materialism: British Natural Philosophy in an Age of Reason* (Princeton University Press, 1970); Theodore M. Brown, 'The College of Physicians and the Acceptance of Iatromechanism in England, 1665–1695', *Bulletin of the History of Medicine*, 44 (1970), 12–30, and Brown, 'From Mechanism to Vitalism in Eighteenth-Century English Physiology', *Journal of the History of Biology*, 7 (1974), 179–216. Steven Shapin discusses the social and institutional factors behind this shift, in 'Social Uses of Science', in *The Ferment of Knowledge: Studies in the Historiography of Eighteenth-Century Science*, ed. G. S. Rousseau and Roy Porter (Cambridge University Press, 1980), 93–139. For changes in medical practice consequent on a rejection of mechanistic theory, see W. F. Bynum, 'Health, Disease and Medical Care', in *Ferment of Knowledge*, ed. Rousseau and Porter, pp. 215–16.
50. On the specificity of Edinburgh medicine see Lawrence, 'Nervous System and Society'.
51. On Whytt, see R. K. French, *Robert Whytt, the Soul and Medicine* (London: Wellcome Institute, 1969), and French, 'Sauvages, Whytt and the Motion of the Heart: Aspects of Eighteenth-Century Animism', *Clio Medica*, 7 (1972), 35–54.
52. Whytt's major work was *An Essay on the Vital and Other Involuntary Motions of Animals* (1751), also presented to the Society. Other work includes 'Account of some Experiments made with Opium on Living and Dying Animals', an attempt to map the relationship of the muscles to the nervous system by demonstrating that opium affects the body via the nerves rather than through the blood, published in the Society's *Essays and Observations, Physical and Literary* (1756). A paper by Whytt in the first volume of *Essays and Observations* (1754), 'Of the Difference between Respiration and Motion of the Heart in Sleeping and Waking Persons', examines the action of the heart during the diminished bodily sensibility of sleep. On the history of the Edinburgh Philosophical Society, see Roger L. Emerson, 'The Philosophical Society of Edinburgh, 1737–1747', *British Journal for the History of Science*, 12 (1979), 154–91, and Emerson, 'The Philosophical Society of Edinburgh 1748–1768', *British Journal for the History of Science*, 14 (1981), 133–76. On Haller, see Andrew Cunningham, 'The Pen and the Sword: Recovering the Disciplinary Identity of Physiology and Anatomy before 1800. I: Old Physiology – the Pen', *Studies in History and Philosophy of Biological and Biomedical Sciences*, 33 (2002), 631–65, and Cunningham, 'Old Anatomy'. On Whytt's dispute with Haller, see also Cunningham, 'Old Anatomy', pp. 67–8.
53. Adam Smith, 'The History of the Ancient Physics', in *Essays on Philosophical Subjects*, pp. 106–17 (pp. 107, 116).

54. Mandeville, 'Remark Y', in *The Fable of the Bees*, vol. 1, p. 250.
55. Smith, *Theory of Moral Sentiments*, p. 276. See also pp. 274–6, 292–3.

4   Enlightenment Legacies and Cultural Radicalism: Physiology and Politics in the 1790s

1. *London Review of English and Foreign Literature*, 7 (1777), 233.
2. *Aberdeen Journal* (1794), quoted in Jessie Dobson, 'John Hunter and the Unfortunate Doctor Dodd', *Journal of the History of Medicine*, 10 (1955), 369–78 (pp. 373–4).
3. *Newcastle Magazine* 1:1 (January 1822), 18–19, and 1:3 (March 1822), 127.
4. See Wendy Moore, *The Knife Man: Blood, Body-Snatching and the Birth of Modern Surgery* (London: Bantam, 2005), pp. 240–2.
5. William Hunter, *Two Introductory Lectures, Delivered by Dr. William Hunter, to his Last Course of Anatomical Lectures, at his Theatre in Windmill Street* (London, 1784), pp. 81, 96.
6. John Hunter, *Treatise on the Blood, Inflammation, and Gunshot Wounds* (London, 1794), p. 78.
7. Hunter, *Treatise on the Blood*, p. 28.
8. John Hunter, 'Proposals for the Recovery of People Apparently Drowned', *Philosophical Transactions of the Royal Society*, 66 (1776), 412–25.
9. Cullen, *Letter*, pp. 2–3.
10. *Transactions of the Royal Humane Society*, 1795.
11. Carolyn Williams discusses the Humane Society as a context for *Frankenstein* in '"Inhumanly brought back to life and misery": Mary Wollstonecraft, *Frankenstein* and the Royal Humane Society', *Women's Writing*, 8:2 (2001), 213–34.
12. On the Magdalen Hospital and sentiment, see Mary Peace, 'The Magdalen Hospital and the Fortunes of Whiggish Sentimentality in Mid-Eighteenth Century Britain', *Eighteenth Century: Theory and Interpretation*, 48:2 (2007), 125–48. Thelwall praises the Magdalen Hospital in a later poem in his volume, comparing its charitable impulses with those of the Humane Society. See John Thelwall, *Poems on Various Subjects* (London, 1787), p. 78.
13. Thelwall, *Poems on Various Subjects*, 'The Seducer', Canto 4, p. 157 (punctuation slightly amended).
14. The strong effects produced by such a mixing of different passions had been described by Hume.
15. Thelwall, *Poems on Various Subjects*, 'The Seducer', Canto 5, p. 172.
16. Thelwall, *Poems on Various Subjects*, 'The Seducer', Canto 5, p. 167.
17. Thelwall, *Poems on Various Subjects*, 'A Dramatic Poem', p. 67.
18. Thelwall, *Poems on Various Subjects*, 'A Dramatic Poem', p. 63.
19. Samuel Horsley, *On the Principle of Vitality in Man, as Described in the Holy Scriptures, and the Difference between True and Apparent Death* (London, 1789), pp. 2, 7.
20. Horsley, *Vitality*, pp. 11, 12.
21. Horsley, *Vitality*, pp. 13, 20.
22. John Hunter, *Lectures on the Principles of Surgery*, in *The Works of John Hunter*, ed. James F. Palmer, 4 vols (London, 1837), vol. 1, p. 213.
23. *The Anti-Jacobin Magazine and Review*, London 1799, p. 1. Edmund Burke, *First Letter on a Regicide Peace*, 1796, in *The Writings and Speeches of Edmund Burke*,

ed. Paul Langford, 12 vols (Oxford: Clarendon Press, 1991), vol. 9, ed. R. B. McDowell, p. 199.

24. Wollstonecraft, *Vindication of the Rights of Men*, p. 9.

25. See John Barrell, *Imagining the King's Death* (Oxford University Press, 2000) and John Whale, *Imagination Under Pressure 1789–1832* (Cambridge University Press, 2000).

26. See, for instance, John Thelwall's *Political Lectures 1* (London, 1794), p. 26, John Thelwall, *Sober Reflections on the Seditious and Inflammatory Letter of the Right Hon. Edmund Burke to a Noble Lord* (London, 1796), pp. 3, 68.

27. Barrell, *Imagining the King's Death*, p. 4.

28. Burke, *First Letter on a Regicide Peace*, p. 190.

29. See John Thelwall, *The Rights of Nature, against the Usurpations of Establishments* (London, 1796), p. 1, Thelwall, *Sober Reflections*, pp. 12, 56, 58.

30. Thelwall, *Rights of Nature*, p. 30.

31. John Thelwall, 'An Essay Towards a Definition of Animal Vitality', repr. in Nicholas Roe, *The Politics of Nature* (Basingstoke: Palgrave, 2002), p. 118.

32. Philip F. Rehbock, 'Transcendental Anatomy', in *Romanticism and the Sciences*, ed. Andrew Cunningham and Nicholas Jardine (Cambridge University Press, 1990), p. 156.

33. See Richardson, *British Romanticism*, and Ruston, *Shelley and Vitality*.

34. For Humphry Davy, see Richardson, *British Romanticism*, p. 52.

35. Richardson, *British Romanticism*, p. xv.

36. Thelwall, 'Animal Vitality', p. 100; Thelwall, *The Peripatetic*, ed. Judith Thompson (Michigan: Wayne State University Press), p. 147.

37. Roe, *Politics of Nature*, p. 94.

38. See Jon Mee, *Romanticism, Enthusiasm, and Regulation* (Oxford University Press, 2003), Paul Keen, *The Crisis of Literature in the 1790s* (Cambridge University Press, 1999) and Andrew McCann, *Cultural Politics in the 1790s* (London: Macmillan, 1999). James Allard's 'John Thelwall and the Politics of Medicine', *European Romantic Review*, 15:1 (2004), 73–87 is rare in exploring the inflection of Thelwall's physiological learning on his politics. The distinction between Thelwall's speech and writings is of course a porous one, given that much of Thelwall's published work was previously delivered in lecture-form. Given that we only have access to his lectures in their published form, they will be discussed under the term 'writings' for the remainder of the chapter.

39. E. P. Thompson, *The Making of the English Working Class* (London: Gollancz, 1963), p. 157.

40. Gregory Claeys, *The Politics of English Jacobinism* (University Park: Pennsylvania State University Press, 1995).

41. Keen, *Crisis of Literature in the 1790s*, pp. 143, 170, 167.

42. Mee, *Romanticism, Enthusiasm, and Regulation*, p. 122.

43. McCann, *Cultural Politics in the 1790s*, pp. 83–106 (p. 84).

44. Burke, *First Letter on a Regicide Peace*, p. 224, and *Second Letter on a Regicide Peace*, p. 292. Energy, activity and communication are regarded with deep suspicion in both texts. Ritson is quoted in Mark Philp, 'The Fragmented Ideology of Reform', in *The French Revolution and British Popular Politics*, ed. Philp (Cambridge University Press, 1991), pp. 50–77 (p. 52).

45. John Thelwall, 'King Chauntclere, or, The Fate of Tyranny', *Politics for the People*, 8 (1794), 102–7 (p. 103).

46. Barrell, *Imagining the King's Death*, pp. 104–8 (p. 106). Michael Scrivener discusses allegory as a primary feature of Jacobin writing in *Seditious Allegories: John Thelwall*

and Jacobin Writing (University Park: Pennsylvania State University Press, 2001). For other discussions of Chaunticlere, see Philp, 'Fragmented Ideology', pp. 70–2, and Marcus Wood, 'William Cobbett, John Thelwall, Radicalism, Racism and Slavery', *Romanticism on the Net*, 15 (1999).

47. Thelwall, *Political Lectures I*, p. viii.
48. Thelwall, 'King Chaunticlere', p. 102.
49. Thelwall, *Sober Reflections*, pp. 10, 11–12.
50. For Galvani, see Wylie, *Young Coleridge*, p. 133.
51. For Thelwall's letter to his wife, see Thompson, *Working Class*, p. 141. *The Tribune*, vol. 3, no. XLVII, Friday 6 November 1795, p. 265.
52. Thelwall, *Political Lectures I*, p. 37; John Thelwall, *The Natural and Constitutional Right of Britons to Annual Parliaments, Universal Suffrage, and the Freedom of Popular Association* (London, 1795), p. 24.
53. *The Tribune*, vol. 1, no. IX, 9 May 1795, p. 224.
54. Thelwall, *Political Lectures I*, pp. 11, 40.
55. Burke, *First Letter on a Regicide Peace*, 1796; 'On the Overtures of Peace', in *The Writings and Speeches of Edmund Burke*, vol. 9, ed. R. B. McDowell (Oxford: Clarendon Press, 1991), pp. 192, 197, 223.
56. Burke, *First Letter on a Regicide Peace*, pp. 230ff, 236.
57. Burke, *First Letter on a Regicide Peace*, pp. 224, 206, 225. Compare Burke, *Letter to a Noble Lord*, where the people are seen as in need of protection from their 'wild and inconsiderate' desires (in *The Writings and Speeches of Edmund Burke*, vol. 9, ed. McDowell, p. 155).
58. Burke, *First Letter on a Regicide Peace*, p. 253. Compare Thelwall, *Rights of Nature*, p. 25: Britain is not the 'mysterious, allegorical thing, which statesmen call Britain'. It is 'the aggregate of the British population'.
59. Thelwall, *Sober Reflections*, pp. 111–12.
60. Thelwall, *Rights of Nature*, p. 3.
61. Thelwall, *Rights of Nature*, pp. 55–9, 87.
62. Thelwall, *Rights of Nature*, pp. 18–19.
63. Thelwall, *Rights of Nature*, p. 71.
64. Thelwall, *Sober Reflections*, p. 81.
65. Thelwall, *Sober Reflections*, p. 114.
66. Thelwall, *Rights of Nature*, pp. 27–8. Italics as in the original.
67. Burke, *First Letter on a Regicide Peace*, p. 190.
68. Thelwall, *Rights of Nature*, p. 28.
69. Thelwall, *Rights of Nature*, p. 39.
70. *The Tribune*, vol. 3, no. XXXIV, 9 October 1795, p. 36.
71. See Barrell, *Imagining the King's Death*, also Olivia Smith, *The Politics of Language 1791–1819* (Oxford: Clarendon Press, 1984), pp. 37–8.
72. Smith, *Politics of Language*, p. x.
73. Scrivener, *Seditious Allegories*.
74. Scrivener, *Seditious Allegories*, p. 10. Compare Philp, 'Fragmented Ideology', p. 69, on the 'innovative character of radical writing in this period'.
75. For attacks on innuendo, see Thelwall, *Political Lectures II*, p. 1, and Thelwall, *Natural and Constitutional Right*, p. 21.
76. Thelwall, *Sober Reflections*, pp. 1, 60–1; Thelwall, *Rights of Nature*, p. 53.
77. *The Tribune*, vol. 2, no. XVI, 29 April 1795, pp. 2–3, and vol. 3, no. XXXIV, 9 October 1795, p. 17.
78. Smith, *Politics of Language*, p. 87.

79. Thelwall, *Sober Reflections*, pp. 5–6.
80. See Smith, *Politics of Language*, p. 26.
81. Thelwall, *Sober Reflections*, pp. 21–32.
82. Thelwall, *Sober Reflections*, pp. 30, 43.
83. Michael Scrivener's phrase, 'hermeneutic aggression', is a useful one in this context. Scrivener, *Seditious Allegories*, p. 43.
84. *The Tribune*, vol. I, no, VII, 25 April 1795, p. 163.
85. John Thelwall, *Poems Written in Close Confinement* (London, 1795). See especially Ode II.
86. Packham, 'Physiology of Political Economy'.
87. See Thelwall, *Political Lectures II*.
88. Quoted in Nicholas Roe, *John Keats and the Culture of Dissent* (Oxford: Clarendon Press, 1997), p. 172.
89. Keen, *Crisis of Literature in the 1790s*, p. 146; Smith, *Politics of Language*, p. 36.
90. Thelwall, *Rights of Nature*, p. 53.

5 Animated Nature: Erasmus Darwin and the Poetry and Politics of Vital Matter, 1789–1803

1. *Analytical Review* (May 1789), repr. in *The Letters and Prose Writings of William Cowper*, ed. James King and Charles Ryskamp, 5 vols (Oxford: Clarendon Press, 1984), vol. 5, pp. 90–9 (pp. 91, 98.)
2. *Analytical Review* (March 1793), repr. in *Letters and Prose Writings*, ed. King and Ryskamp, vol. 5, pp. 100–8 (p. 106). '*Verbosum curiosa felicitas*' is 'the studied felicity of words'.
3. For context and an early draft, see Cowper's letter to William Hayley, June 1792, in *Letters and Prose Writings*, ed. King and Ryskamp, vol. 4, pp. 106–8.
4. See Desmond King-Hele, *Erasmus Darwin: A Life of Unequalled Achievement* (London: De la Mare, 1999), pp. 237–8.
5. For detailed discussion of the reception of Darwin, see Norton Garfinkle, 'Science and Religion in England, 1790–1800: The Critical Response to the Work of Erasmus Darwin', *Journal of the History of Ideas*, 16:3 (1955), 376–88, and Julia List, 'Erasmus Darwin's Beautification of the Sublime: Materialism, Religion and the Reception of *The Economy of Vegetation* in the Early 1790s', *Journal for Eighteenth-Century Studies*, 32:3 (2009), 389–405.
6. *Critical Review*, 6 (October 1792), pp. 162–71 (p. 169); *Monthly Review*, 11 (June 1793), pp. 182–7 (p. 186).
7. King-Hele, *Erasmus Darwin*, p. 265.
8. Advertisement, *The Loves of the Plants*. Parts of the argument in this section are drawn from my article, 'The Science and Poetry of Animation'.
9. King-Hele, *Erasmus Darwin*, pp. 96, 201.
10. Erasmus Darwin, *Zoonomia; or, the Laws of Organic Life*, 2 vols (London, 1794–96), vol. 1, pp. 1–2; emphasis as in the original.
11. Darwin, *Zoonomia*, vol. 1, p. 32.
12. Darwin, *Zoonomia*, vol. 1, pp. 10, 66.
13. Darwin, *Zoonomia*, vol. 1, p. 109.
14. For Thomas Brown's more informed accusation of Darwin's materialism, see Richardson, *British Romanticism*, p. 15. Richardson's helpful account of Darwin differs from that given here in its primary focus on his brain theory.

15. Richardson, *British Romanticism*, pp. 29–30, 69, 192 n. 58.
16. Darwin, *Zoonomia*, vol. 1, pp. 32–3, 101–4.
17. Darwin, *Zoonomia*, vol. 1, pp. 104–5.
18. Darwin, *Zoonomia*, vol. 1, pp. 102, 105.
19. Darwin, *Zoonomia*, vol. 1, p. 1.
20. George Campbell, *The Philosophy of Rhetoric* (1776, repr. Carbondale: Southern Illinois University Press, 1963), pp. 73, 75, 53.
21. Thomas Reid, *An Inquiry into the Human Mind on the Principles of Common Sense* (1764), in *The Works of Thomas Reid*, ed. William Hamilton, 2 vols (Edinburgh, 1863, repr. Bristol: Thoemmes, 1994), I, p. 201.
22. *Monthly Review*, 43:2 (February 1804), 113–27 (pp. 124–5); *Edinburgh Review*, II (July 1803), 491–506 (pp. 498–9). For the *Edinburgh Review*'s political orientation, see Philip Connell, *Romanticism, Economics and the Question of 'Culture'* (Oxford University Press, 2001), pp. 93ff.
23. *Monthly Review*, 43:2 (February 1804), pp. 122, 118, 124.
24. *Quarterly Review*, 12 (1814), pp. 60–90 (p. 71); Review of 'Collected Works of the Late Dr. Sayers', *Quarterly Review*, 35 (1827), pp. 175–220 (pp. 198–9). I am grateful to Professor David Fairer for drawing my attention to these references.
25. For a fuller exposition of this argument, see my 'The Science and Poetry of Animation'. For Wordsworth's rejection of Darwin, see *The Fenwick Notes of William Wordsworth* (1993), p. 170, quoted in King-Hele, *Erasmus Darwin*, p. 265.
26. See 'The Futility of Criticism', *Weekly Magazine*, 3 (12 January 1760), in *Collected Works of Oliver Goldsmith*, ed. Arthur Friedman, 5 vols (Oxford: Clarendon Press, 1966), III, pp. 51–3 (p. 53); 'On Poetry, as Distinguished from Other Writing', in *The Works of Oliver Goldsmith*, 4 vols (London: George Bell, 1892), I, p. 357.
27. See Joseph Addison, 'An Essay on Virgil's *Georgics*', in *Eighteenth-Century Critical Essays*, ed. Scott Elledge, 2 vols (Ithaca, NY: Cornell University Press, 1961), I, pp. 1–8 (p. 2).
28. Anna Seward, *Memoirs of the Life of Dr. Darwin* (Philadelphia, 1804), pp. 94–5. For women and botanical writing in the late eighteenth century, see Samantha George, *Botany, Sexuality, and Women's Writing, 1760–1830* (Manchester University Press, 2007).
29. Goldsmith, 'Futility of Criticism', p. 53.
30. Erasmus Darwin, *The Botanic Garden, containing The Loves of the Plants* (London, 1789), pp. 132, vi, 83, 40.
31. Darwin, *The Botanic Garden*, p. vi.
32. Darwin, *Zoonomia*, vol. 1, pp. 102, 105, 107.
33. Darwin, *The Botanic Garden*, pp. 39–44.
34. According to Philip Connell, an established opposition between 'literature, aesthetics, and feeling, on the one hand; and science, utility and reason' was to become intransigent by the late 1820s. See *Romanticism, Economics and the Question of 'Culture'*, p. 11. For the argument that Wordsworth's 'Preface' was a response to the claims made for science by Humphry Davy in his Royal Institution Lectures, see Roger Sharrock, 'The Chemist and the Poet; Sir Humphry Davy and the Preface to the *Lyrical Ballads*', *Notes and Records of the Royal Society of London*, 17 (1962), pp. 57–76.
35. Christopher Lawrence offers a rather different reading of Wordsworth in 'The Power and the Glory'.
36. Darwin, *Zoonomia*, vol. 1, pp. 130–1.

37. *The Spectator*, no. 421 (July 1712), in *The Spectator*, ed. Smith, III, pp. 304–7.
38. See Lawrence, 'The Power and the Glory', pp. 222–3.
39. *The Anti-Jacobin, or Weekly Examiner*, published from 1797 to 1798 and edited by William Gifford, is not to be confused with *The Anti-Jacobin Review and Magazine*, a successor publication, edited by John Gifford, and which appeared from 1798 to 1819.
40. See the *British Critic*, 23 (February 1804), 169–74.
41. On Priestley, Bentham, and the social and political agenda of radical philosophy from the 1770s, see Simon Schaffer, 'States of Mind: Enlightenment and Natural Philosophy', in *The Languages of Psyche*, ed. G. S. Rousseau (Berkeley: University of California Press, 1990). On Beddoes, see also Lawrence, 'The Power and the Glory', and Brian Dolan, 'Conservative Politicians, Radical Philosophers, and the Aerial Remedy for the Diseases of Civilisation', *History of the Human Sciences*, 15:2 (2002), 35–54.
42. Anon., *The Golden Age: A Poetical Epistle* (London, 1794), footnote, p. 6.
43. Anon., *The Golden Age*, p. 8.
44. *The Anti-Jacobin*, April 1798, pp. 162–7, and June 1798, pp. 415–19.
45. *The Anti-Jacobin*, April 1798, pp. 164–5.
46. *The Anti-Jacobin*, June 1798, p. 417.
47. *The Anti-Jacobin*, April 1798, p. 163.
48. King-Hele, *Erasmus Darwin*, p. 246. On Priestley and aerial improvements, see Schaffer, 'States of Mind', and Dolan, 'Conservative Politicians'.
49. For French chemists' participation in revolutionary activities, such as the collection of saltpetre for gunpowder, as well as Burke's influential hostility to chemistry, see Maurice Crosland, 'The Image of Science as a Threat: Burke versus Priestley and the "Philosophic Revolution"', *British Journal for the History of Science*, 20 (1987), 277–307 (pp. 285–7).
50. *The Anti-Jacobin*, April 1798, p. 163. Hannah More warned of the dangers of German writing, including drama, in *Strictures on the Modern System of Female Education* (1799). See Jones, ed., *Women in the Eighteenth Century*, pp. 132–4.
51. Thomas Mathias, *The Pursuits of Literature, or What You Will: A Satirical Poem in Dialogue* (part 1) (London, 1794), pp. 14–15.
52. Although mostly by Canning, some lines of *New Morality* are attributed to his collaborator, George Ellis.
53. See for instance Markman Ellis' discussion in *The Politics of Sensibility* (Cambridge University Press, 1996), pp. 192–7.
54. *New Morality, The Anti-Jacobin*, July 1798, p. 627, ll. 119–24.
55. *New Morality*, ll. 224, 52–3, 451–2, 434.
56. For more discussion of Darwin's influence on Coleridge and others, including evidence for links between *Zoonomia* and Wordsworth's *Lyrical Ballads*, see King-Hele, *Erasmus Darwin*.
57. The phrase 'creative anthology' is Marcus Wood's: see *Radical Satire and Print Culture, 1790–1822* (Oxford: Clarendon Press, 1994), p. 88.
58. *Pig's Meat*, 2 vols (London, 1795), vol. 2, p. 124; vol. 1, p. 33.
59. See Keen, *Crisis of Literature in the 1790s*.
60. See Scrivener, *Seditious Allegories*. The fable of Chaunticlere, discussed in the previous chapter, is one example of a more complex Jacobin text.
61. Erasmus Darwin, *The Economy of Vegetation*, Canto II, ll. 385–8, 391–4, repr. in *Politics for the People*, vol. I, no. 12, p. 165. The canto and line references given in Eaton's journal are incorrect.

62. Darwin, *Economy of Vegetation*, Canto IV, ll. 389–92.
63. Thelwall, *Sober Reflections*, p. 116.
64. See *The Tribune*, vol. 1, p. 91.
65. Thelwall's earlier publication, *The Peripatetic*, rooted in accounts of his wanderings around London and its immediate environs, is one example of his ability to make telling political use of even the most mundane of his experiences.

6 Animation and Vitality in Women's Writing of the 1790s

1. Shelley, *Frankenstein*, p. 97.
2. Wollstonecraft, *Mary*, p. 46. Mary is here quoting Edward Young's *Night Thoughts*.
3. Mary Wollstonecraft, *The Wrongs of Woman: or, Maria*, in *The Works of Mary Wollstonecraft*, ed. Janet Todd and Marilyn Butler, 7 vols (London: Pickering and Chatto), vol. 1. p. 134.
4. Mary Shelley's Introduction to the 1831 *Frankenstein*, in *Frankenstein*, p. 178.
5. Edmund Burke, *Reflections on the Revolution in France*, ed. Conor Cruise O'Brien (London: Penguin, 1968), p. 195.
6. See Shelley's Introduction to the 1831 *Frankenstein* for the full account.
7. See Marilyn Butler's introduction to her Oxford University Press edition of *Frankenstein* (1993) for a convenient route into this work. Other scientific contexts for Shelley's text have also been mooted: see Laura Crouch, 'Davy, *A Discourse*: A Possible Scientific Source of Frankenstein', *Keats-Shelley Journal*, 27 (1978), 35–44, and Jan Golinski's *Science as Public Culture: Chemistry and Enlightenment in Britain, 1760–1820* (Cambridge University Press, 1992), which links *Frankenstein* not directly to vitalist science but to the new chemical discoveries of the late eighteenth century. For a discussion of the materialism of both Mary and Percy Shelley, see Paul Hamilton, *Metaromanticism: Aesthetics, Literature, Theory* (University of Chicago Press, 2003), pp. 139–55.
8. For the Abernethy/Lawrence debate, see Ruston, *Shelley and Vitality*, and Butler's edition of *Frankenstein*, Appendix C.
9. See Erasmus Darwin, *The Temple of Nature, or The Origin of Society* (London, 1803), Additional Note 1 on Spontaneous Generation, p. 7. Shelley misremembers the experiment as on 'vermicelli'.
10. Mary Robinson, *A Letter to the Women of England* (London, 1799), p. 44.
11. Mary Ann Radcliffe, *The Female Advocate* (London, 1799), pp. 40–1.
12. Radcliffe, *Female Advocate*, pp. 165, 169, 174.
13. Golinski, *Science as Public Culture*. For women and science, see pp. 76 and 102; also Londa Schiebinger, *The Mind has no Sex? Women in the Origins of Modern Science* (Cambridge, MA: Harvard University Press, 1989), chs 1–2.
14. Golinski, *Science as Public Culture*, p. 6.
15. Golinski, *Science as Public Culture*, p. 51.
16. Both quoted by Golinski, *Science as Public Culture*, pp. 81 and 82.
17. Memoir, prefaced to John Thelwall, *Poems, Chiefly Written in Retirement* (Hereford, 1801), p. xxiii.
18. See Crosland, 'The Image of Science as a Threat'.
19. Simon Schaffer and Steven Shapin, *Leviathan and the Air Pump* (Princeton University Press, 1985). Golinski, *Science as Public Culture*, p. 102.
20. *Monthly Review*, December 1796; November 1795.
21. *Monthly Review*, May 1794.

22. *The Anti-Jacobin Review and Magazine*, 1:1, July 1798.
23. *Analytical Review*, 8 October 1790, repr. in *The Works of Mary Wollstonecraft*, ed. Todd and Butler, vol. 7, pp. 293–300.
24. For one recent study of Romantic representations of natural force, see Ted Underwood, *The Work of the Sun: Literature, Science, and Political Economy, 1760–1860* (Basingstoke: Palgrave, 2005).
25. For the limitations of such an approach, see Ellis, *The Politics of Sensibility*, pp. 10ff.
26. Eliza Fenwick, *Secresy, or The Ruin on the Rock*, ed. Isobel Grundy (Ontario: Broadview, 1998).
27. Nicola Watson, *Revolution and the Form of the British Novel* (Oxford: Clarendon Press, 1994), pp. 40–4.
28. Julia M. Wright, '"I am ill fitted": Conflicts of Genre in Eliza Fenwick's *Secresy*', in *Romanticism, History, and the Possibilities of Genre*, ed. Tilottama Rajan and Julia M. Wright (Cambridge University Press, 1998), pp. 149–75.
29. Terry Castle, 'Eliza Fenwick and Eighteenth-Century Women's Writing', *The London Review of Books*, 17:4 (1995), 18–19.
30. Fenwick, *Secresy*, pp. 94, 103, 66.
31. Fenwick, *Secresy*, pp. 44, 75.
32. Fenwick, *Secresy*, p. 43.
33. Fenwick, *Secresy*, p. 76.
34. Fenwick, *Secresy*, p. 64.
35. Fenwick, *Secresy*, p. 344.
36. Fenwick, *Secresy*, pp. 249, 123.
37. Fenwick, *Secresy*, p. 254.
38. Johnson, *Equivocal Beings*, pp. 46–69.
39. Wollstonecraft, *Mary*, pp. 68, 73.
40. Wollstonecraft, *Mary*, p. 46.
41. *Analytical Review*, XII (January 1792), repr. in *The Works of Mary Wollstonecraft*, ed. Todd and Butler, vol. 7, pp. 411–13 (p. 412).
42. Gary Kelly, *Revolutionary Feminism: The Mind and Career of Mary Wollstonecraft* (Basingstoke: Macmillan, 1992), pp. 40–54. See also his introduction to his Oxford World's Classics edition of *Mary and The Wrongs of Woman* (Oxford University Press, 1976), pp. vii–xxi, esp. p. xi.
43. See Taylor, *Mary Wollstonecraft and the Feminist Imagination*, p. 62.
44. Edward Young, *Conjectures on Original Composition* (1759), repr. in *The Sublime: A Reader in British Eighteenth-Century Aesthetic Theory*, ed. Andrew Ashfield and Peter de Bolla (Cambridge University Press, 1996), p. 113.
45. Simon Schaffer, 'Genius in Romantic Natural Philosophy', in *Romanticism and the Sciences*, ed. Andrew Cunningham and Nicholas Jardine (Cambridge University Press, 1990), pp. 82–98 (p. 92).
46. Taylor, *Mary Wollstonecraft and the Feminist Imagination*, p. 36.
47. Wollstonecraft, *Mary*, p. 46.
48. Wollstonecraft, *Mary*, p. 53.
49. Wollstonecraft, *Mary*, pp. 58, 59–61, 63.
50. Taylor, *Mary Wollstonecraft and the Feminist Imagination*, pp. 97ff.
51. *Collected Letters of Mary Wollstonecraft*, ed. Ralph M. Wardle (Ithaca, NY: Cornell University Press, 1979), p. 149.
52. Wollstonecraft, *Mary*, p. 46. The quotation is from Edward Young's *Night Thoughts*, the italics are as in the original.

53. William Duff, *An Essay on Original Genius* (1767), quoted in Taylor, *Mary Wollstonecraft and the Feminist Imagination*, pp. 61–2.
54. Wollstonecraft, *The Wrongs of Woman*, Author's Preface, p. 83.
55. Wollstonecraft, *The Wrongs of Woman*, p. 163. On Pennington, see Ellis, *The Politics of Sensibility*, pp. 29–33.
56. Wollstonecraft, *The Wrongs of Woman*, pp. 120, 87, 88, 167.
57. Wollstonecraft, *The Wrongs of Woman*, pp. 138–9.
58. Wollstonecraft, *The Wrongs of Woman*, pp. 157, 158.
59. Wollstonecraft, *The Wrongs of Woman*, pp. 133, 138.
60. Wollstonecraft, *The Wrongs of Woman*, p. 93.
61. Wollstonecraft, *The Wrongs of Woman*, p. 96.
62. Wollstonecraft, *The Wrongs of Woman*, pp. 90, 131, 173.
63. Wollstonecraft, *The Wrongs of Woman*, p. 92.
64. McCann, *Cultural Politics in the 1790s*, pp. 154–5.
65. Wollstonecraft, *The Wrongs of Woman*, pp. 144, 163.
66. Ellis, *The Politics of Sensibility*, pp. 5–8.
67. Harriet Guest, *Small Change: Women, Learning, Patriotism, 1750–1810* (University of Chicago Press, 2000), p. 310.
68. The secondary literature on the physiology of sensibility is extensive, but includes Rousseau, 'Nerves, Spirits and Fibres', Van Sant, *Eighteenth-Century Sensibility and the Novel*, and Mullan, *Sentiment and Sociability*.
69. Wollstonecraft, *The Wrongs of Woman*, p. 163.
70. Johnson, *Equivocal Beings*, p. 67.
71. Wollstonecraft, *The Wrongs of Woman*, pp. 176–7.
72. Wollstonecraft, *The Wrongs of Woman*, p. 141.

Conclusion: Eighteenth-Century Vitalism, Romanticism, Literature and the Disciplines

1. David Fairer's *Organising Poetry: The Coleridge Circle, 1790–1798* (Oxford University Press, 2009) also challenges an outmoded opposition of eighteenth-century mechanism and Romantic organicism – an opposition which, as Fairer points out, goes back to M. H. Abrams' *The Mirror and The Lamp*. Fairer argues that eighteenth-century thinking in a number of areas, from personal identity to history to poetic tradition, could arguably be described as 'organic', and what he characterises as an 'eighteenth-century organic' needs to be understood on its own terms, distinct from the Romantic organic of, for instance, the mature Coleridge. See also Stephen Gaukroger, *The Collapse of Mechanism and the Rise of Sensibility: Science and the Shaping of Modernity 1680–1760* (Oxford University Press, 2010).
2. Charles Armstrong, *Romantic Organicism* (Basingstoke: Palgrave, 2003), p. 5. For Tilottama Rajan's discussion of the organic analogies which were frequently deployed in writings of many kinds between 1780 and 1830, see her 'Organicism', *English Studies in Canada*, 30:4 (2004), 46–50 (p. 50). See also Gigante, *Life: Organic Form and Romanticism*.
3. See, for instance, James Engell, *The Creative Imagination: Enlightenment to Romanticism* (Cambridge, MA: Harvard University Press, 1981).
4. Terry Eagleton's *The Ideology of the Aesthetic* (Oxford: Blackwell, 1990) is an obvious theoretical landmark here, but for representative work in the Romantic

period see, for instance, Nigel Leask, *The Politics of Imagination in Coleridge's Critical Thought* (Basingstoke: Macmillan, 1988), or Ellis, *The Politics of Sensibility*. For further discussion of this point, see the opening section of my 'Feigning Fictions'.

5. Frederick Burwick, *Approaches to Organic Form* (Dordrecht: D. Reidel, 1987), pp. ix–x.
6. Rajan, 'Organicism', p. 48.
7. Rajan, 'Organicism', and Tilottama Rajan, 'The Unavowable Community of Idealism: Coleridge and the Life Sciences', *European Romantic Review*, 14:4 (2003), 395–416.
8. Armstrong, *Romantic Organicism*, p. 9
9. Hamilton, *Metaromanticism*, pp. 139–55.
10. For the Romantic century, see Susan Wolfson, '50–50? Phone a Friend? Speculating on a Romantic Century, 1750–1850', *European Romantic Review*, 11:1 (2000), 1–11. Also on the question of periodisation, see Michael McKeon, 'Recent Studies in the Restoration and Eighteenth Century', *Studies in English Literature*, 45:3 (2005), 707–82, Clifford Siskin, 'Personification and Community: Literary Change in the Mid and Late Eighteenth Century', *Eighteenth-Century Studies*, 15:4 (1982), 371–401, and Miriam L. Wallace, 'Enlightened Romanticism or Romantic Enlightenment?', in *Enlightening Romanticism, Romancing Enlightenment*, ed. Miriam L. Wallace (Farnham: Ashgate, 2009), pp. 1–20.
11. Cunningham, 'Old Physiology', p. 649. Richard Sha, *Perverse Romanticism: Aesthetics and Sexuality in Britain, 1750–1832* (Baltimore: Johns Hopkins University Press, 2009) also finds Cunningham's distinction between physiology pre- and post-1800 a significant one: see ch. 2.
12. Cunningham, 'Old Physiology', p. 654.
13. Cunningham, 'Old Anatomy', p. 58.
14. Cunningham, 'Old Anatomy', p. 65.
15. Cunningham, 'Old Anatomy', pp. 68–70.
16. Cunningham, 'Old Physiology', p. 659.
17. Cunningham, 'Old Physiology', p. 637.
18. In her survey of evidence for demarcating the 'Romantic century', for instance, Susan Wolfson pays little attention to scientific thought.
19. Patricia Spacks also singles out the 1790s for special consideration in debates over periodisation in literary history, though her particular concern is with the novel. See Patricia M. Spacks, 'How We See: The 1790s', in *Enlightening Romanticism, Romancing Enlightenment*, ed. Miriam L. Wallace (Farnham: Ashgate, 2009), pp. 179–88.
20. Mary Poovey, 'The Model System of Contemporary Literary Criticism', *Critical Inquiry*, 27:3 (2001), 408–38 (p. 412).
21. Poovey, 'Model System', p. 418. See also Clifford Siskin, *The Historicity of Romantic Discourse* (Oxford University Press, 1988), Siskin, *The Work of Writing* (Baltimore: Johns Hopkins University Press, 1998) and Connell, *Romanticism, Economics and the Question of 'Culture'*.
22. Poovey, 'Model System', pp. 419ff. Coleridge's organicism is also central to Fairer and Armstrong's studies.

Bibliography

Allard, James, 'John Thelwall and the Politics of Medicine', *European Romantic Review*, 15:1 (2004), 73–87

Anon., *The Golden Age: A Poetical Epistle* (London, 1794)

Arbuthnot, John, *Gnothi Seauton: Know Yourself* (London, 1734), repr. in *Literature and Science 1660–1834*, 8 vols (London: Pickering and Chatto, 2003), vol. 2, ed. Clark Lawlor and Akihito Suzuki

Armstrong, Charles, *Romantic Organicism* (Basingstoke: Palgrave, 2003)

Armstrong, John, *The Œconomy of Love: A Poetical Essay* (London: T. Cooper, 1736)

—— *The Art of Preserving Health* (London, 1744), repr. in *Literature and Science 1660–1834*, 8 vols (London: Pickering and Chatto, 2003), vol. 2, ed. Clark Lawlor and Akihito Suzuki

Ashfield, Andrew, and Peter de Bolla, *The Sublime: A Reader in British Eighteenth-Century Aesthetic Theory* (Cambridge University Press, 1996)

Barfoot, Michael, 'Hume and the Culture of Science in the Early Eighteenth Century', in *Studies in the Philosophy of the Scottish Enlightenment*, ed. M. A. Stewart (Oxford: Clarendon Press, 1990), pp. 151–90

Barker-Benfield, G. J., *The Culture of Sensibility* (University of Chicago Press, 1992)

Barrell, John, 'The Public Prospect and the Private View: The Politics of Taste in Eighteenth-Century Britain', in *Reading Landscape: Country-City-Capital*, ed. Simon Pugh (Manchester University Press, 1990), pp. 19–40

—— *Imagining the King's Death* (Oxford University Press, 2000)

Battersby, Christine, 'Hume, Newton and the "Hill Called Difficulty"', in *Philosophers of the Enlightenment*, ed. S. C. Brown (Brighton: Harvester Press, 1979), pp. 31–55

Blacklock, Thomas, *Poems on Several Occasions* (Glasgow, 1746)

—— *Poems on Several Occasions* (Edinburgh, 1754)

—— *Poems by Mr. Thomas Blacklock* (London, 1756)

—— entry on 'Blind', *Encyclopædia Britannica*, vol. 2, 2nd edn (Edinburgh, 1778)

—— *Poems by the Late Reverend Dr. Thomas Blacklock* (Edinburgh, 1793)

Brown, Theodore M., 'The College of Physicians and the Acceptance of Iatromechanism in England, 1665–1695', *Bulletin of the History of Medicine*, 44 (1970), 12–30

—— 'From Mechanism to Vitalism in Eighteenth-Century English Physiology', *Journal of the History of Biology*, 7 (1974), 179–216

Buickerood, James G., 'The Natural History of the Understanding: Locke and the Rise of Facultative Logic in the Eighteenth Century', *History and Philosophy of Logic*, 6 (1985), 157–90

Burke, Edmund, *Reflections on the Revolution in France*, ed. Conor Cruise O'Brien (London: Penguin, 1968)

—— *The Writings and Speeches of Edmund Burke*, ed. Paul Langford, 12 vols (Oxford: Clarendon Press, 1991)

Burwick, Frederick, ed., *Approaches to Organic Form* (Dordrecht: D. Reidel, 1987)

Butler, Marilyn, and Janet Todd, eds, *The Works of Mary Wollstonecraft*, 7 vols (London: Pickering, 1989)

Bynum, W. F., 'Health, Disease and Medical Care', in *Ferment of Knowledge*, ed. Rousseau and Porter, pp. 215–16

Campbell, Archibald, *An Enquiry into the Original of Moral Virtue* (London, 1734)

Campbell, George, *The Philosophy of Rhetoric* (1776, repr. Carbondale: Southern Illinois University Press, 1963)

Campbell, R. H., and Andrew S. Skinner, eds, *The Origins and Nature of the Scottish Enlightenment* (Edinburgh: John Donald, 1982)

Campbell, T. D., 'Francis Hutcheson: "Father" of the Scottish Enlightenment', in *The Origins and Nature of the Scottish Enlightenment*, ed. R. H. Campbell and Andrew S. Skinner (Edinburgh: John Donald, 1982), pp. 167–85

Cant, Ronald G., 'Origins of the Enlightenment in Scotland: The Universities', in *The Origins and Nature of the Scottish Enlightenment*, ed. R. H. Campbell and Andrew S. Skinner (Edinburgh: John Donald, 1982), pp. 42–64

Capaldi, Nicholas, *David Hume: The Newtonian Philosopher* (Boston: Twayne, 1975)

Castle, Terry, 'Eliza Fenwick and Eighteenth-Century Women's Writing', *The London Review of Books*, 23:2 (1995)

Christie, John R. R., 'The Origins and Development of the Scottish Scientific Community, 1680–1760', *History of Science*, 12 (1974), 122–41

Claeys, Gregory, *The Politics of English Jacobinism* (University Park: Pennsylvania State University Press, 1995)

Coleridge, Samuel Taylor, *Letters of Samuel Taylor Coleridge*, ed. E. L. Griggs (Oxford: Clarendon Press, 1956), vol. 1

—— *Theory of Life*, in *The Collected Works of Samuel Taylor Coleridge*, vol. 11, *Shorter Works and Fragments*, ed. H. J. Jackson and J. R. de J. Jackson (London: Routledge, 1995), pp. 481–557

Connell, Philip, *Romanticism, Economics and the Question of 'Culture'* (Oxford University Press, 2001)

Cooper, Anthony Ashley, Third Earl of Shaftesbury, *Characteristicks of Men, Manners, Opinions, Times* (London, 1732)

Copley, Stephen, and David Fairer, '*An Essay on Man* and the Polite Reader', in *Pope: New Contexts*, ed. David Fairer (Hemel Hempstead: Harvester Wheatsheaf, 1990)

Crosland, Maurice, 'The Image of Science as a Threat: Burke versus Priestley and the "Philosophic Revolution"', *British Journal for the History of Science*, 20 (1987), 277–307

Crouch, Laura, 'Davy, *A Discourse*: A Possible Scientific Source of Frankenstein', *Keats-Shelley Journal*, 27 (1978), 35–44

Cullen, William, *A Letter to Lord Cathcart, President of the Board of Police in Scotland, Concerning the Recovery of Persons Drowned and Seemingly Dead* (Edinburgh, 1776)

—— *Clinical Lectures Delivered in the Years 1765 and 1766* (London, 1797)

—— *The Works of William Cullen*, ed. John Thomson, 2 vols (London and Edinburgh, 1827)

Cunningham, Andrew, 'The Pen and the Sword: Recovering the Disciplinary Identity of Physiology and Anatomy before 1800. I: Old Physiology – the Pen', *Studies in History and Philosophy of Biological and Biomedical Sciences*, 33 (2002), 631–5

—— 'The Pen and the Sword: Recovering the Disciplinary Identity of Physiology and Anatomy before 1800. II: Old Anatomy – the Sword', *Studies in History and Philosophy of Biological and Biomedical Sciences*, 34 (2003), 51–76

Cunningham, Andrew, and Nicholas Jardine, eds, *Romanticism and the Sciences* (Cambridge University Press, 1990)

Darwin, Erasmus, *The Botanic Garden, containing The Loves of the Plants* (London, 1789)

—— *Zoonomia; or, the Laws of Organic Life*, 2 vols (London, 1794–96)

—— *The Temple of Nature, or The Origin of Society* (London, 1803)

De Almeida, Hermione, *Romantic Medicine and John Keats* (Oxford University Press, 1991)

De Bolla, Peter, 'The Visibility of Visuality', in *Vision in Context: Historical and Contemporary Perspectives on Sight*, ed. Teresa Brennan and Martin Jay (London: Routledge, 1996), pp. 65–79

Desmond, Adrian, *The Politics of Evolution: Morphology, Medicine and Reform in Radical London* (Chicago University Press, 1989)

Deutsch, Helen, and Felicity Nussbaum, eds, *Defects: Engendering the Modern Body* (Ann Arbor: University of Michigan Press, 2000)

Dobson, Jessie, 'John Hunter and the Unfortunate Doctor Dodd', *Journal of the History of Medicine*, 10 (1955), 369–78

Doig, A., J. P. S. Ferguson, I. A. Milne and R. Passmore, eds, *William Cullen and the Eighteenth-Century Medical World* (Edinburgh University Press, 1993)

Dolan, Brian, 'Conservative Politicians, Radical Philosophers, and the Aerial Remedy for the Diseases of Civilisation', *History of the Human Sciences*, 15:2 (2002), 35–54

Donovan, A. L., *Philosophical Chemistry in the Scottish Enlightenment: The Doctrines and Discoveries of William Cullen and Joseph Black* (Edinburgh University Press, 1975)

Duchesneau, François, 'Vitalism in Late Eighteenth-Century Physiology: The Cases of Barthez, Blumenbach and John Hunter', in *William Hunter and the Eighteenth-Century Medical World*, ed. W. F. Bynum and Roy Porter (Cambridge University Press, 1985)

Eagleton, Terry, *The Ideology of the Aesthetic* (Oxford: Blackwell, 1990)

[Edinburgh Philosophical Society], *Essays and Observations, Physical and Literary* (1754, 1756)

Elledge, Scott, ed., *Eighteenth-Century Critical Essays*, 2 vols (Ithaca, NY: Cornell University Press, 1961)

Ellis, Markman, *The Politics of Sensibility* (Cambridge University Press, 1996)

Emerson, Roger L., 'The Philosophical Society of Edinburgh, 1737–1747', *British Journal for the History of Science*, 12 (1979), 154–91

—— 'The Philosophical Society of Edinburgh 1748–1768', *British Journal for the History of Science*, 14 (1981), 133–76

—— 'Natural Philosophy and the Problem of the Scottish Enlightenment', *Studies on Voltaire and the Eighteenth Century*, 242 (1986), 243–91

—— 'Science and Moral Philosophy in the Scottish Enlightenment', in *Studies in the Philosophy of the Scottish Enlightenment*, ed. Stewart, pp. 11–36

Engell, James, *The Creative Imagination: Enlightenment to Romanticism* (Cambridge, MA: Harvard University Press, 1981)

Fairer, David, ed., *Pope: New Contexts* (Hemel Hempstead: Harvester Wheatsheaf, 1990)

—— *Organising Poetry: The Coleridge Circle, 1790–1798* (Oxford University Press, 2009)

Fenwick, Eliza, *Secresy, or The Ruin on the Rock*, ed. Isobel Grundy (Ontario: Broadview, 1998)

Ferguson, Frances, 'Organic Form and its Consequences', in *Land, Nation, and Culture 1740–1840: Thinking the Republic of Taste*, ed. Peter de Bolla, Nigel Leask and David Simpson (Basingstoke: Palgrave, 2005)

Finkelstein, Andrea, *Harmony and the Balance: An Intellectual History of Seventeenth-Century English Economic Thought* (Ann Arbor: University of Michigan Press, 2000)

Fordyce, David, *Elements of Moral Philosophy* (London, 1754)

Fox, Christopher, *Locke and the Scriblerians: Identity and Consciousness in Early Eighteenth-Century Britain* (Berkeley: University of California Press, 1988)

French, R. K., *Robert Whytt, the Soul and Medicine* (London: Wellcome Institute, 1969)
—— 'Sauvages, Whytt and the Motion of the Heart: Aspects of Eighteenth-Century Animism', *Clio Medica*, 7 (1972), 35–54
Furniss, Tom, *Edmund Burke's Aesthetic Ideology: Language, Gender and Political Economy in Revolution* (Cambridge University Press, 1993)
Gabbey, Alan, 'Newton, Active Powers, and the Mechanical Philosophy', in *The Cambridge Companion to Newton*, ed. I. Bernard Cohen and George E. Smith (Cambridge University Press, 2002)
Garfinkle, Norton, 'Science and Religion in England, 1790–1800: The Critical Response to the Work of Erasmus Darwin', *Journal of the History of Ideas*, 16:3 (1955), 376–88
Gaukroger, Stephen, *The Collapse of Mechanism and the Rise of Sensibility: Science and the Shaping of Modernity 1680–1760* (Oxford University Press, 2010)
George, Samantha, *Botany, Sexuality, and Women's Writing, 1760–1830* (Manchester University Press, 2007)
Gigante, Denise, *Life: Organic Form and Romanticism* (New Haven and London: Yale University Press, 2009)
Goldsmith, Oliver, *The Works of Oliver Goldsmith*, 4 vols (London: George Bell, 1892)
—— *Collected Works of Oliver Goldsmith*, ed. Arthur Friedman, 5 vols (Oxford: Clarendon Press, 1966)
Golinski, Jan, *Science as Public Culture: Chemistry and Enlightenment in Britain, 1760–1820* (Cambridge University Press, 1992)
Guerrini, Anita, 'James Keill, George Cheyne, and Newtonian Physiology, 1690–1740', *Journal of the History of Biology*, 18 (1985), 247–66
—— 'The Tory Newtonians: Gregory, Pitcairne, and their Circle', *Journal of British Studies*, 25 (1986), 288–311
Guest, Harriet, *Small Change: Women, Learning, Patriotism, 1750–1810* (University of Chicago Press, 2000)
Haigh, Elizabeth L., 'Vitalism, the Soul, and Sensibility: The Physiology of Theophile Bordeu', *Journal of the History of Medicine and Allied Sciences*, 31 (1976), 30–41
Hamilton, Paul, *Metaromanticism: Aesthetics, Literature, Theory* (University of Chicago Press, 2003)
Hankins, Thomas L., *Science and the Enlightenment* (Cambridge University Press, 1985)
Heimann, P. M., '"Nature is a Perpetual Worker": Newton's Aether and Eighteenth-Century Natural Philosophy', *Ambix*, 20 (1973), 1–25
—— 'Voluntarism and Immanence: Conceptions of Nature in Eighteenth-Century Thought', *Journal of the History of Ideas*, 39 (1978), 271–83
Heimann, P. M., and J. E. McGuire, 'Newtonian Forces and Lockean Powers: Concepts of Matter in Eighteenth-Century Thought', *Historical Studies in the Physical Sciences*, 3 (1971), 233–306
Hirschman, Albert O., *The Passions and the Interests: Political Arguments for Capitalism before its Triumph* (Princeton University Press, 1977)
Holmes, Richard, *The Age of Wonder* (London: Harper, 2008)
Hont, Istvan, and Michael Ignatieff, *Wealth and Virtue: The Shaping of Political Economy in the Scottish Enlightenment* (Cambridge University Press, 1983)
Horsley, Samuel, *On the Principle of Vitality in Man, as Described in the Holy Scriptures, and the Difference between True and Apparent Death* (London, 1789)
Hume, David, *The Letters of David Hume*, ed. J. Y. T. Greig, 2 vols (Oxford: Clarendon Press, 1932)
—— *Enquiries Concerning Human Understanding and the Principles of Morals*, ed. L. A. Selby-Bigge and P. H. Nidditch (Oxford: Clarendon Press, 1975)

—— *A Treatise of Human Nature*, ed. L. A. Selby-Bigge (Oxford: Clarendon Press, 1978)

—— *Essays: Moral, Political and Literary*, ed. Eugene Miller (Indianapolis: Liberty Fund, 1985)

—— *Four Dissertations* (London, 1757, repr. Bristol: Thoemmes Press, 1995)

Hunter, John, 'Proposals for the Recovery of People Apparently Drowned', *Philosophical Transactions of the Royal Society*, 66 (1776), 412–25

—— *Treatise on the Blood, Inflammation, and Gunshot Wounds* (London, 1794)

—— *Lectures on the Principles of Surgery*, in *The Works of John Hunter*, ed. James F. Palmer, 4 vols (London, 1837), vol. 1

Hunter, William, *Two Introductory Lectures, Delivered by Dr. William Hunter, to his Last Course of Anatomical Lectures, at his Theatre in Windmill Street* (London, 1784)

Hutcheson, Francis, *A Short Introduction to Moral Philosophy, in Three Books* (Glasgow, 1747)

—— *A System of Moral Philosophy* (Glasgow, 1755)

Jacyna, L. S., 'Immanence or Transcendence: Theories of Life and Organization in Britain, 1790–1835', *Isis*, 74 (1983), 311–29

James, Susan, *Passion and Action: The Emotions in Seventeenth-Century Philosophy* (Oxford: Clarendon Press, 1997)

Johnson, Claudia L., *Equivocal Beings: Politics, Gender and Sentimentality in the 1790s* (Chicago University Press, 1995)

Johnson, Samuel, *The Rambler*, no. 168 (26 October 1751), 377–9

Jones, Chris, *Radical Sensibility* (London: Routledge, 1993)

Jones, Peter, ed., *Philosophy and Science in the Scottish Enlightenment* (Edinburgh: John Donald, 1988)

Jones, Vivien, ed., *Women in the Eighteenth Century* (London: Routledge, 1990)

Jordanova, Ludmilla, ed., *Languages of Nature: Critical Essays on Science and Literature* (London: Free Association, 1986)

Keen, Paul, *The Crisis of Literature in the 1790s* (Cambridge University Press, 1999)

Kelly, Gary, *Revolutionary Feminism: The Mind and Career of Mary Wollstonecraft* (Basingstoke: Macmillan, 1992)

King, James, and Charles Ryskamp, *The Letters and Prose Writings of William Cowper*, 5 vols (Oxford: Clarendon Press, 1984)

King-Hele, Desmond, *Erasmus Darwin: A Life of Unequalled Achievement* (London: De la Mare, 1999)

Koelb, Clayton, *The Revivifying Word: Literature, Philosophy, and the Theory of Life in Europe's Romantic Age* (Rochester, NY: Camden House, 2008)

Kroll, Richard, *The Material Word: Literate Culture in the Restoration and Early Eighteenth Century* (Baltimore: Johns Hopkins University Press, 1991)

Larrissy, Edward, 'The Celtic Bard of Romanticism: Blindness and Second Sight', *Romanticism*, 5.1 (1999), 43–57

Lawrence, Christopher, 'The Nervous System and Society in the Scottish Enlightenment', in *Natural Order: Historical Studies of Scientific Culture*, ed. Barry Barnes and Steven Shapin (London: Sage, 1979), pp. 19–40

—— 'The Power and the Glory: Humphry Davy and Romanticism', in *Romanticism and the Sciences*, ed. Andrew Cunningham and Nicholas Jardine (Cambridge University Press, 1990), pp. 213–27

Leechman, William, 'Preface: Giving some Account of the Life, Writings, & Character of the Author', in *A System of Moral Philosophy*, by Francis Hutcheson (Glasgow, 1755), pp. i–xlviii

Levere, Trevor H., *Poetry Realized in Nature: Samuel Taylor Coleridge and Early Nineteenth Century Science* (Cambridge University Press, 1981)

—— 'Coleridge and the Sciences', in *Romanticism and the Sciences*, ed. Andrew Cunningham and Nicholas Jardine (Cambridge University Press, 1990), pp. 295–306

List, Julia, 'Erasmus Darwin's Beautification of the Sublime: Materialism, Religion and the Reception of *The Economy of Vegetation* in the Early 1790s', *Journal for Eighteenth-Century Studies*, 32:3 (2009), 389–405

Locke, John, *An Essay Concerning Human Understanding*, ed. Peter H. Nidditch (Oxford: Clarendon Press, 1975)

Lowry, S. Todd, 'The Archaeology of the Circulation Concept in Economic Theory', *Journal of the History of Ideas*, 35 (1974), 429–44

Mackenzie, Henry, 'Some Account of the Life and Writings of Dr. Blacklock', in *Poems by the Late Reverend Dr. Thomas Blacklock* (Edinburgh, 1793)

Mandeville, Bernard, *The Fable of the Bees: Or, Private Vices, Publick Benefits*, ed. F. B. Kaye, 2 vols (Oxford: Clarendon Press, 1924)

Marshall, David, *The Figure of Theatre: Shaftesbury, Defoe, Adam Smith and George Eliot* (New York: Columbia University Press, 1986)

Mathias, Thomas, *The Pursuits of Literature, or What You Will: A Satirical Poem in Dialogue* (part 1) (London, 1794)

McCann, Andrew, *Cultural Politics in the 1790s* (London: Macmillan, 1999)

McKeon, Michael, 'Recent Studies in the Restoration and Eighteenth Century', *Studies in English Literature*, 45:3 (2005), 707–82

—— 'Biography, Fiction, and the Emergence of "Identity" in Eighteenth-Century Britain', in *Writing Lives: Biography and Texuality, Identity and Representation in Early Modern England*, ed. Kevin Sharpe and Steven N. Zwicker (Oxford University Press, 2008), pp. 339–55

Mee, Jon, *Romanticism, Enthusiasm, and Regulation* (Oxford University Press, 2003)

Moore, Wendy, *The Knife Man: Blood, Body-Snatching and the Birth of Modern Surgery* (London: Bantam, 2005)

Moravia, Sergio, 'From "Homme Machine" to "Homme Sensible": Changing Eighteenth-Century Models of Man's Image', *Journal of the History of Ideas*, 39 (1978), 45–60

Morrell, J. B., 'The University of Edinburgh in the Late Eighteenth Century', *Isis*, 62 (1971), 158–71

Mossner, Ernest Campbell, 'Hume's Epistle to Dr. Arbuthnot, 1734: The Biographical Significance', *Huntington Library Quarterly*, 7 (1944), 135–52

Mullan, John, *Sentiment and Sociability* (Oxford: Clarendon Press, 1990)

Newton, Isaac, *Opticks, or A Treatise of the Reflections, Refractions, Inflections, and Colours of Light* (London, 1730, repr. London: Bell and Sons, 1931)

—— *Mathematical Principles*, trans. Andrew Motte, rev. Florian Cajori (Berkeley: University of California Press, 1934)

Nussbaum, Felicity, *The Autobiographical Subject: Gender and Ideology in Eighteenth-Century England* (Baltimore: Johns Hopkins University Press, 1989)

—— *The Limits of the Human: Fictions of Anomaly, Race and Gender in the Long Eighteenth Century* (Cambridge University Press, 2003)

Packham, Catherine, 'The Physiology of Political Economy: Vitalism and Adam Smith's *Wealth of Nations*', *Journal of the History of Ideas*, 63:3 (2002), 465–81

—— 'The Science and Poetry of Animation: Personification, Analogy, and Erasmus Darwin's *Loves of the Plants*', *Romanticism*, 10:2 (2004), 191–208

—— 'Feigning Fictions: Imagination, Hypothesis, and Philosophical Writing in the Scottish Enlightenment', *Eighteenth Century: Theory and Interpretation*, 44:2 (2007), 149–71

—— 'Disability and Sympathetic Sociability in Enlightenment Scotland: The Case of Thomas Blacklock', *British Journal for Eighteenth-Century Studies*, 30:3 (2007), 423–38

Peace, Mary, 'The Magdalen Hospital and the Fortunes of Whiggish Sentimentality in Mid-Eighteenth Century Britain', *Eighteenth Century: Theory and Interpretation*, 48:2 (2007), 125–48

Philp, Mark, 'The Fragmented Ideology of Reform', in *The French Revolution and British Popular Politics*, ed. Philp (Cambridge University Press, 1991), pp. 50–77

Pocock, J. G. A., *Virtue, Commerce, and History: Essays on Political Thought and History, Chiefly in the Eighteenth Century* (Cambridge University Press, 1985)

Poovey, Mary, *A History of the Modern Fact: Problems of Knowledge in the Sciences of Wealth and Society* (University of Chicago Press, 1998)

—— 'The Model System of Contemporary Literary Criticism', *Critical Inquiry*, 27:3 (2001), 408–38

Pope, Alexander, *The Poems of Alexander Pope*, ed. John Butt (1963, repr. London: Routledge, 1989)

Porter, Roy, *Flesh in the Age of Reason* (London: Penguin, 2004)

Quinlan, Sean M., 'Apparent Death in Eighteenth-Century France and England', *French History*, 9 (1995), 27–47

Radcliffe, Mary Ann, *The Female Advocate* (London, 1799)

Rajan, Tilottama, 'The Unavowable Community of Idealism: Coleridge and the Life Sciences', *European Romantic Review*, 14:4 (2003), 395–416

—— 'Organicism', *English Studies in Canada*, 30:4 (2004), 46–50

Raynor, David R., 'Hume's Abstract of Adam Smith's *Theory of Moral Sentiments*', *Journal of the History of Philosophy*, 22 (1984), 51–79

Rehbock, Philip F., 'Transcendental Anatomy', in *Romanticism and the Sciences*, ed. Andrew Cunningham and Nicholas Jardine (Cambridge University Press, 1990)

Reid, Thomas, *An Inquiry into the Human Mind on the Principles of Common Sense* (1764), in *The Works of Thomas Reid*, ed. William Hamilton, 2 vols (Edinburgh, 1863, repr. Bristol: Thoemmes, 1994), vol. 1

Reill, Peter Hanns, 'The Legacy of the "Scientific Revolution": Science and the Enlightenment', in *The Cambridge History of Science*, vol. 4, *Eighteenth-Century Science*, ed. Roy Porter (Cambridge University Press, 2003), pp. 23–43

—— *Vitalizing Nature in the Enlightenment* (Berkeley: University of California Press, 2005)

—— 'Eighteenth-Century Uses of Vitalism in Constructing the Human Sciences', in *Biology and Ideology from Descartes to Dawkins*, ed. Denis R. Alexander and Ronald L. Numbers (University of Chicago Press, 2010), pp. 61–371

Richardson, Alan, *British Romanticism and the Science of the Mind* (Cambridge University Press, 2001)

Risse, Guenter B., 'Doctor William Cullen, Physician, Edinburgh: A Consultative Practice in the Eighteenth Century', *Bulletin of the History of Medicine*, 48 (1974), 338–51

Robinson, Mary, *A Letter to the Women of England* (London, 1799)

Roe, Nicholas, *John Keats and the Culture of Dissent* (Oxford: Clarendon Press, 1997)

—— ed., *Samuel Taylor Coleridge and the Sciences of Life* (Oxford University Press, 2001)

—— *The Politics of Nature: William Wordsworth and Some Contemporaries* (Basingstoke: Palgrave, 2002)

Roe, Shirley, 'The Life Sciences', in *The Cambridge History of Science*, vol. 4, *Eighteenth-Century Science*, ed. Roy Porter (Cambridge University Press, 2003), pp. 397–416

Rousseau, George, ed., *The Languages of Psyche: Mind and Body in Enlightenment Thought* (Berkeley: University of California Press, 1990)
—— 'Nerves, Spirits and Fibres: Towards the Origins of Sensibility', in *Nervous Acts*, by George Rousseau (New York: Palgrave, 2004)
Rousseau, George, and Roy Porter, eds, *The Ferment of Knowledge: Studies in the Historiography of Eighteenth-Century Science* (Cambridge University Press, 1980)
Ruston, Sharon, *Shelley and Vitality* (Basingstoke: Palgrave, 2005)
Schaffer, Simon, 'Genius in Romantic Natural Philosophy', in *Romanticism and the Sciences*, ed. Andrew Cunningham and Nicholas Jardine (Cambridge University Press, 1990)
—— 'States of Mind: Enlightenment and Natural Philosophy', in *The Languages of Psyche*, ed. G. S. Rousseau (Berkeley: University of California Press, 1990)
Schaffer, Simon, and Steven Shapin, *Leviathan and the Air Pump* (Princeton University Press, 1985)
Schiebinger, Londa, *The Mind has no Sex? Women in the Origins of Modern Science* (Cambridge, MA: Harvard University Press, 1989)
Schofield, Robert E., *Mechanism and Materialism: British Natural Philosophy in an Age of Reason* (Princeton University Press, 1970)
Scrivener, Michael, *Seditious Allegories: John Thelwall and Jacobin Writing* (University Park: Pennsylvania State University Press, 2001)
Sekora, John, *Luxury: The Concept in Western Thought* (Baltimore: Johns Hopkins University Press, 1977)
Seller, William, 'Memoir of the Life and Writings of Robert Whytt M. D.', *Transactions of the Royal Society of Edinburgh*, 23 (1864), 99–131
Seward, Anna, *Memoirs of the Life of Dr. Darwin* (Philadelphia, 1804)
Sha, Richard, *Perverse Romanticism: Aesthetics and Sexuality in Britain, 1750–1832* (Baltimore: Johns Hopkins University Press, 2009)
Shapin, Steven, 'Social Uses of Science', in *Ferment of Knowledge*, ed. Rousseau and Porter, pp. 93–139
Sharrock, Roger, 'The Chemist and the Poet: Sir Humphry Davy and the Preface to the *Lyrical Ballads*', *Notes and Records of the Royal Society of London*, 17 (1962), 57–76
Shelley, Mary, *Frankenstein, or the Modern Prometheus*, ed. Nora Crook (London: Pickering and Chatto, 1996)
Shepherd, Christine M., 'Newtonianism in Scottish Universities in the Seventeenth Century', in *Origins and Nature of Scottish Enlightenment*, ed. Campbell and Skinner, pp. 65–85
Shuttleton, David, '"A Modest Examination": John Arbuthnot and the Scottish Newtonians', *British Journal for Eighteenth-Century Studies*, 18:1 (1995), 47–62
Siskin, Clifford, 'Personification and Community: Literary Change in the Mid and Late Eighteenth Century', *Eighteenth-Century Studies*, 15:4 (1982), 371–401
—— *The Historicity of Romantic Discourse* (Oxford University Press, 1988)
—— *The Work of Writing* (Baltimore: Johns Hopkins University Press, 1998)
Smith, Adam, *An Inquiry into the Nature and Causes of the Wealth of Nations* (Oxford University Press, 1976)
—— *The Theory of Moral Sentiments*, ed. D. D. Raphael and A. L. MacFie (Oxford University Press, 1976)
—— *Essays on Philosophical Subjects*, ed. W. P. D. Wightman and J. C. Bryce (Oxford: Clarendon Press, 1980)
—— *Lectures on Rhetoric and Belles Lettres*, ed. J. C. Bryce (Oxford University Press, 1983)

—— *The Correspondence of Adam Smith*, ed. Ernest Campbell Mossner and Ian Simpson Ross (Oxford: Clarendon, 1987)

Smith, Olivia, *The Politics of Language 1791–1819* (Oxford: Clarendon Press, 1984)

Spacks, Patricia M., *Imagining a Self: Autobiography and Novel in Eighteenth-Century England* (Cambridge, MA: Harvard University Press, 1976)

—— 'How We See: The 1790s', in *Enlightening Romanticism, Romancing Enlightenment*, ed. Miriam L. Wallace (Farnham: Ashgate, 2009), pp. 179–88

The Spectator, ed. Gregory Smith, 4 vols (London: Dent, 1963)

Spence, Joseph, 'An Account of the Life, Character, and Poems of the Author', in *Poems by Mr. Thomas Blacklock* (London, 1756)

Stafford, Barbara Maria, *Body Criticism: Imaging the Unseen in Enlightenment Art and Medicine* (Cambridge, MA: MIT, 1991)

Stewart, M. A., ed., *Studies in the Philosophy of the Scottish Enlightenment*, (Oxford: Clarendon Press, 1990)

Stott, Rosalie, 'Health and Virtue: Or, How to Keep out of Harm's Way: Lectures on Pathology and Therapeutics by William Cullen c. 1770', *Medical History*, 31 (1987), 123–42

Taylor, Barbara, *Mary Wollstonecraft and the Feminist Imagination* (Cambridge University Press, 2003)

Thelwall, John, *Poems on Various Subjects* (London, 1787)

—— 'An Essay Towards a Definition of Animal Vitality' (1793), repr. in Nicholas Roe, *The Politics of Nature* (Basingstoke: Palgrave, 2002)

—— *The Peripatetic* (1793), ed. Judith Thompson (Michigan: Wayne State University Press, 2001)

—— *Political Lectures I and II* (London, 1794)

—— *The Natural and Constitutional Right of Britons to Annual Parliaments, Universal Suffrage, and the Freedom of Popular Association* (London, 1795)

—— *Poems Written in Close Confinement* (London, 1795)

—— *The Rights of Nature, against the Usurpations of Establishments* (London, 1796)

—— *Sober Reflections on the Seditious and Inflammatory Letter of the Right Hon. Edmund Burke to a Noble Lord* (London, 1796)

—— *Poems, Chiefly Written in Retirement* (Hereford, 1801)

—— *Letter to Henry Cline, Esq* (1810)

Thompson, E. P., *The Making of the English Working Class* (London: Gollancz, 1963)

Todd, Dennis, *Imagining Monsters: Miscreations of the Self in Eighteenth-Century England* (University of Chicago Press, 1995)

Transactions of the Royal Humane Society (1795)

The Tribune, Consisting Chiefly of the Political Lectures of J. Thelwall, 3 vols (London, 1794–96)

Trumpener, Katie, *Bardic Nationalism: The Romantic Novel and the British Empire* (Princeton University Press, 1997)

Underwood, Ted, *The Work of the Sun: Literature, Science, and Political Economy, 1760–1860* (Basingstoke: Palgrave, 2005)

Van Sant, Ann Jessie, *Eighteenth-Century Sensibility and the Novel* (Cambridge University Press, 1993)

Wahrmann, Dror, *The Making of the Modern Self: Identity and Culture in Eighteenth-Century England* (New Haven: Yale University Press, 2004)

Wallace, Miriam L., 'Enlightened Romanticism or Romantic Enlightenment?', in *Enlightening Romanticism, Romancing Enlightenment*, ed. Miriam L. Wallace (Farnham: Ashgate, 2009), pp. 1–20

Wallace, William A., *Causality and Scientific Explanation*, 2 vols (Ann Arbor: University of Michigan Press, 1974)

Watson, Nicola, *Revolution and the Form of the British Novel* (Oxford: Clarendon Press, 1994)

Whale, John, 'Romantic Attacks: Pope and the Spirit of Language', in *Pope: New Contexts*, ed. David Fairer (Hemel Hempstead: Harvester Wheatsheaf, 1990), pp. 153–68

—— 'Death in the Face of Nature: Self, Society and Body in Wollstonecraft's *Letters Written in Sweden, Norway and Denmark*', *Romanticism*, 1:2 (1995), 177–92

—— *Imagination Under Pressure 1789–1832* (Cambridge University Press, 2000)

Whytt, Robert, 'Observations on the Sensibility and Irritability of the Parts of Men and Other Animals', in *Physiological Essays*, 2nd edn (Edinburgh, 1761)

—— *Observations on the Nature, Causes, and Cure of those Disorders which have been commonly called Nervous, Hypochondriac, or Hysteric: To which are prefixed some Remarks on the Sympathy of the Nerves*, 2nd edn (Edinburgh, 1765)

—— 'Essay on the Vital and Other Involuntary Motions of Animals' (1751), in *The Works of Robert Whytt* (Edinburgh, 1768)

Williams, Carolyn, '"Inhumanly brought back to life and misery": Mary Wollstonecraft, *Frankenstein* and the Royal Humane Society', *Women's Writing*, 8:2 (2001), 213–34

Wolfson, Susan, '50–50? Phone a Friend? Speculating on a Romantic Century, 1750–1850', *European Romantic Review*, 11:1 (2000), 1–11

Wollstonecraft, Mary, *Collected Letters of Mary Wollstonecraft*, ed. Ralph M. Wardle (Ithaca, NY: Cornell University Press, 1979)

—— *Mary, a Fiction* and *The Wrongs of Woman: or, Maria*, in *The Works of Mary Wollstonecraft*, ed. Janet Todd and Marilyn Butler, 7 vols (London: Pickering and Chatto, 1989), vol. 1

—— *Letters Written During a Short Residence in Sweden, Norway, and Denmark*, in *The Works of Mary Wollstonecraft*, ed. Janet Todd and Marilyn Butler, 7 vols (London: Pickering and Chatto, 1989), vol. 6

—— *A Vindication of the Rights of Men* and *A Vindication of the Rights of Woman*, in *The Works of Mary Wollstonecraft*, ed. Janet Todd and Marilyn Butler, 7 vols (London: Pickering and Chatto, 1989), vol. 5

Wood, Marcus, *Radical Satire and Print Culture, 1790–1822* (Oxford: Clarendon Press, 1994)

—— 'William Cobbett, John Thelwall, Radicalism, Racism and Slavery', *Romanticism on the Net*, 15 (1999)

Wood, Paul, 'Science and the Pursuit of Virtue in the Aberdeen Enlightenment', in *Origins and Nature of the Scottish Enlightenment*, ed. Campbell and Skinner, pp. 127–49

—— 'Science and the Aberdeen Enlightenment', in *Philosophy and Science in the Scottish Enlightenment*, ed. Jones, pp. 39–66

Wright, John P., *The Sceptical Realism of David Hume* (Manchester University Press, 1983)

—— 'Metaphysics and Physiology: Mind, Body, and the Animal Economy in Eighteenth-Century Scotland', in *Studies in the Philosophy of the Scottish Enlightenment*, ed. M. A. Stewart (Oxford: Clarendon Press, 1990), pp. 251–301

Wright, Julia M., '"I am ill fitted": Conflicts of Genre in Eliza Fenwick's *Secresy*', in *Romanticism, History, and the Possibilities of Genre*, ed. Tilottama Rajan and Julia M. Wright (Cambridge University Press, 1998)

Wylie, Ian, *Young Coleridge and the Philosophers of Nature* (Oxford: Clarendon Press, 1989)

Yolton, John, *Thinking Matter: Materialism in Eighteenth-Century Britain* (Oxford: Blackwell, 1983)

Index

Abernethy, John, 7–8, 127, 151, 158, 162, 177–8, 210
Addison, Joseph, 72–3, 153, 154, 157, 208
Analytical Review, 147, 185, 193
animism, 2, 6, 211
Anti-Jacobin, or Weekly Examiner, 159
Anti-Jacobin Review and Magazine, 185
Arbuthnot, John, 12, 25–31, 32, 42–4, 48, 57
Armstrong, Charles, 208, 209
Armstrong, John, 12, 29, 31, 32, 37–44, 57
Austen, Jane, 190

Barrell, John, 125, 130–1, 136
Barthez, Paul-Joseph, 6
Battersby, Christine, 90–1
Beddoes, Thomas, 160, 161, 166
Bentham, Jeremy, 160
Blacklock, Thomas, 12, 29, 32, 43–51, 52, 54, 56, 60
Blair, Hugh, 155
Blumenbach, Johann Friedrich, 6
body politic, 19, 98–9
 in Adam Smith, 101, 102–3, 106–8
 in Burke and Thelwall, 134
 in Mandeville, 106–7
Boerhaave, Hermann, 5–6
Bordeu, Théophile, 6, 104
British Critic, 159
Burke, Edmund, 45, 113, 122–6, 129, 132, 133–9, 143, 167, 168, 176, 183
 Letter on a Regicide Peace, 122, 125, 133–4
 Letter to a Noble Lord, 137, 138–9
 Reflections on the Revolution in France, 166
Butler, Joseph, 16
Byron, Lord, 176, 178

Campbell, Archibald, 62
Campbell, George, 153

Canning, George, 122, 159, 161–8, 169–70, 174
Castle, Terry, 187
Cheselden, William, 74–6
Cheyne, George, 30
Coleridge, Samuel Taylor, 1, 9–10, 149, 158, 160, 169–70, 209, 212, 215
 Theory of Life, 209
Collins, Anthony, 16
Collins, William, 155
Cowper, William, 147–8, 149
Critical Review, 148
Cullen, William, 6, 19–21, 104, 115, 117, 206
Cunningham, Andrew, 210–13

Darwin, Erasmus, 3, 4, 13, 120, 147–74, 177, 178, 182, 183, 186, 189, 192, 197, 209, 210, 213
 Botanic Garden, The, 147–8, 149–51, 154, 156, 161, 162, 173, 178; *see also The Economy of Vegetation*; *The Loves of the Plants*
 Economy of Vegetation, The, 148, 149, 154, 166, 172
 Loves of the Plants, The, 147–8, 149–50, 152–7, 159, 160
 Temple of Nature, The, 13, 148, 153, 154, 159, 161, 165, 166, 178
 Zoonomia, or the Laws of Organic Life, 149, 150–3, 156–7, 160, 162, 166
Davy, Humphry, 10, 127, 160, 182
De Bolla, Peter, 66
Defoe, Daniel, 98–9, 101, 102
Descartes, René, 11, 17, 104
Dodd, William, 111–12, 114, 115, 116, 119
Dryden, John, 154, 170
Duff, William, 197

Eaton, Daniel Isaac, 130, 137, 158, 170–4
Edgeworth, Maria, 115

Edinburgh Medical School, 5–6, 18,
    29, 88, 104–5; *see also* William
    Cullen; William Porterfield;
    Robert Whytt
Edinburgh Philosophical Society, 5, 7, 105
Edinburgh Review, 153
Ellis, Markman, 203

Fairer, David, 235 n. 1
Fenwick, Eliza, 11, 13, 181–2, 185, 186
    Secresy, 179, 186–93, 203
Fielding, Henry, 142
Fordyce, David, 33–5, 38, 62–3
Furniss, Tom, 88, 94, 97

Galvani, Luigi, 132, 151, 212
Gibbon, Edward, 113
Gillray, James, 122, 123, 124, 135–6,
    162, 163, 167, 168
Godwin, William, 122, 129, 162, 170,
    173, 200
Goldsmith, Oliver, 154, 155, 170
Golinski, Jan, 182–3
Gray, Thomas, 155, 171
Grundy, Isobel, 187
Guest, Harriet, 203

Habermas, Jürgen, 141
Haller, Albrecht von, 5, 9, 105, 211–12
Hamilton, Paul, 210
Harvey, William, 98
Hawes, William, *see* Humane Society
Hirschman, Albert O., 87
Hobbes, Thomas, 61
Horsley, Samuel, 119–21, 184
Humane Society, 115–19, 120, 121
Hume, David, 5, 6, 7, 12, 18–19, 32, 35–6,
    43–4, 46–7, 48, 52, 60, 65, 86, 142
    *Enquiry Concerning Human
        Understanding*, 17, 64
    *Enquiry Concerning the Principles of
        Morals*, 64
    'Of the Passions', 91
    Of Refinement in the Arts, 90
    Treatise of Human Nature, 32, 46, 64,
        89–91, 94–6; introduction, 97
    autobiographical writings, 94–6
    education, 91
    early illness, 95–6
    last illness, 103

and mechanism, 90–1
and Newtonian natural philosophy,
    98, 225 n. 22
on the passions, 89, 91, 125
on philosophical labour, 94–7
on the 'springs of action', 89–91
Hunter, John, 4, 7, 12, 38, 112–15, 116,
    117, 119–21, 126–7, 132, 142,
    162, 177, 184, 185, 204, 206, 209
    Lectures on the Principles of Surgery,
        113, 120
    'Proposals for the Recovery of People
        Apparently Drowned', 114
    *Treatise on the Blood, Inflammation, and
        Gunshot Wounds*, 113, 127
    see also under vital principle
Hunter, William, 113, 115, 121, 142
Hutcheson, Francis, 31–2, 34, 35, 36, 41,
    54, 61, 62
Hutton, Charles, 112, 119
Hutton, James, 184
hydrostatics, 91

Johnson, Claudia, 192, 204
Johnson, Joseph, 147, 182, 185; *see also*
    Analytical Review
Johnson, Samuel, 8, 98, 111, 170

Kames, (Henry Home) Lord, 5, 155
Keats, John, 149
Keen, Paul, 129, 130, 132, 140, 142, 171
Kelly, Gary, 193, 195
Knight, Richard Payne, 165

Lavoisier, Antoine, 166, 201
Lawrence, Christopher, 142
Lawrence, William, 7–8, 127, 151, 158,
    177, 184, 210
Leechman, William, 32–3, 34
Locke, John, 10, 11, 20–1, 26, 27–8, 45,
    53, 86, 93–4, 207, 208, 210
    Essay on Human Understanding, 5,
        13–16, 46, 75, 89
    Second Treatise of Government, 88

Mackenzie, Henry, 32, 43, 45
Magdalen Hospital, 111, 116
Mandeville, Bernard, 61, 89, 106–7
materialism, 7, 10, 12, 26, 151–2, 158,
    192, 233 n. 7, 210, 211

Mathias, Thomas, 167, 169
McCann, Andrew, 129, 140, 203
mechanism, 1–2, 5, 9, 10, 12, 15, 27,
    29–30, 33–4, 150, 157, 207, 211
mechanistic physiology, 6, 27, 29, 104,
    120
  in England, 104–5
  Erasmus Darwin's attack on, 150
  William Hunter's attack on, 113
Mee, Jon, 129, 133
Milton, John, 132, 175–6, 196
Monthly Review, 148, 153, 183
Montpellier vitalists, 104

Naturphilosophie, 209
Newton, Isaac, 1–2, 4–5, 10, 17–19, 69,
    207; *see also* Newtonianism
Newtonianism, 2, 5, 32, 121
  in Hume, 90–1

organic, 211, 214, 215; *see also* organicism
  in Romanticism, 154
organicism, 2, 9, 207–10, 212–13, 215–16

Paine, Tom, 129, 139
Parson, John Weddell, 193
Physiocrats, *see* François Quesnay
Pitt, William, 122, 135–6
Place, Francis, 142
Pocock, J. G. A., 87, 97
Polidori, John, 177
Poovey, Mary, 213–15
Pope, Alexander, 2–3, 4, 11, 25–6,
    147–8, 169
Porter, Roy, 10
Porterfield, William, 104, 105
Price, Richard, 182
Priestley, Joseph, 7, 119, 121, 160, 166,
    167, 182–3, 201, 210

Quarterly Review, 154
Quesnay, François, 102

Radcliffe, Mary Ann, 13, 175–6, 179–81,
    186
Rajan, Tilottama, 208, 209
Ramazzini, Bernardino, 85
Rehbock, Philip, 126
Reid, Thomas, 153
Reill, Peter Hanns, 2

Richardson, Alan, 8, 128, 151
Richardson, Samuel, 187
Ritson, Joseph, 129
Rivington, Francis and Charles, 159
Robinson, Mary, 13, 175–6, 178, 187, 191
Roe, Nicholas, 8, 128
Romanticism, 8–10, 154, 169–70, 207–16
Rowe, Nicholas, 175–6
Rowlandson, Thomas, 162, 167
Rousseau, George, 10, 53
Rousseau, Jean-Jacques, 168, 187, 188,
    193, 197, 202, 204–5
Ruston, Sharon, 8–9

Sauvages, Boissier de, 6
Schaffer, Simon, 183, 194
Schrivener, Michael, 137, 172
'science of man', 28, 31–7, 61, 63
sensibility, 10, 11–12, 46–8, 52–82,
    187–8, 203–6, 208
  in Adam Smith, 60–7, 204
  in Erasmus Darwin, 151, 161, 168–9
  in physiology of von Haller and
      Whytt, 9, 105
  in Théophile Bordeu, 104
  and vitalism, 53–4
  in Wollstonecraft, 52–3, 55–9, 186,
      193, 195–6, 198–200, 203–6, 207
Seward, Anna, 154
Shapin, Steven, 183
Shelley, Mary, 175–8, 192
  Frankenstein, 8, 76, 95, 115, 175–8,
      186, 191, 206
Shelley, Percy, 8–9, 14, 177–8, 212
Smellie, William, 185
Smith, Adam, 6–7, 19, 55, 141
  'History of Ancient Physics', 68–70,
      106
  'History of Astronomy', 68–9, 86–7
  *An Inquiry into the Nature and Causes
      of the Wealth of Nations*, 12, 83–8,
      91–3, 99–103, 106–8:
    body in, 92–3, 98–103, 106–8, *see
      also* body politic; on colonial
      trade, 100–1; labour and
      labouring subject in, 83–8,
      91–3, 106–7; on labour wages,
      91–2; on mercantilism, 101–2; on
      monopoly, 101; self-betterment in,
      93, 101, 103, 106–7; on taxes, 101

Smith, Adam – *continued*
  Lectures on Rhetoric and Belles Lettres,
    225 n. 21
  The Theory of Moral Sentiments, 12, 36,
    53, 54–6, 59–67, 69–74, 107, 204,
    205: stoicism in, 107, 141; *see also*
    sensibility
  attends anatomy lectures, 113
  correspondence with Hume, 103–4
  on François Quesnay, 102
  on philosophical labour, 86–7, 93–4,
    96–7
  on philosophical method, 93–4
Smith, Olivia, 137–8, 140, 142
Smollett, Tobias, 142
Southey, Robert, 149, 154, 158
The Spectator, 20–1, 45, 67, 72–3
Spence, Joseph, 45–7
Spence, Thomas, 170, 173
Stahl, Georg, 6
Stewart, Dugald, 86
Stewart, John, 5
Swift, Jonathan, 11

Taylor, Barbara, 197
Thelwall, John, 1, 12, 108, 116–19,
    122–43, 158, 162, 170, 172–4,
    178, 182, 183, 186, 213
  *Essay towards a Definition of Animal
    Vitality*, 126–8, 132, 142
  King Chaunticlere: The Fate of Tyranny,
    130–1
  'On the Origin of Sensation', 127
  The Peripatetic, 122, 126, 128, 143
  Poems, Chiefly Written in Retirement, 111
  Poems on Various Subjects, 116–19
  Poems Written in Close Confinement,
    141, 143
  Political Lectures, 133
  The Rights of Nature, 125–6, 134, 136,
    137, 139
  Sober Reflections, 135, 138–9, 172–3
  The Tribune, 137, 140, 141
Thompson, E. P., 128, 141
Todd, Dennis, 11
Tooke, John Horne, 162
Trumpener, Katie, 46

vital forces, 2, 5, 7, 20, 29, 30, 40, 104,
    135, 148, 185, 190, 192, 194, 204

vital principle, 2, 7, 11, 16, 18, 20, 36,
    38, 55, 60, 105, 115, 126–7, 132,
    184–5, 189, 191, 193, 206, 209,
    212
  in Adam Smith, 106
  Erasmus Darwin on, 151
  John Hunter's search for, 113–14, 115,
    120, 127
  John Thelwall on, 126–7
  in resuscitation, 19–20, 115
  in Robert Whytt, 6, 18–19
  Samuel Horsley's sermon on,
    119–20
  William Cullen's definition of, 19, 115
vitalism, 1–2, 3–8, 9–11, 18–19, 27, 31,
    34, 183–6, 192, 207–16; *see also*
    vital forces; vital principle; vitalist
    physiology
vitalist physiology, 5–7, 11–12, 16,
    29–30, 32, 43, 104–5, 108, 120,
    141, 194, 203–4, 206, 211–12
  in Edinburgh Medical School, 29,
    104–55
  in Erasmus Darwin, 151, 156–8, 174
  gothic vitalism, 188
  organic vitalism, 209
  Romantic vitalism, 1
vitality, language of, 2, 4, 9, 12–13,
    119–21, 128, 142–3, 152, 178,
    184–5, 187–8, 190, 207–8, 209,
    215–16
Volta, Alessandro, 151

Walpole, Horace, 147–8
Watson, Nicola, 187
Whale, John, 58
Whytt, Robert, 5–7, 9, 18–19, 104–6,
    113, 115, 120, 204
Williams, Raymond, 125
Willis, Thomas, 16, 53, 212
Wollstonecraft, Mary, 12, 13, 27, 55,
    74, 122, 175–6, 179, 181–4, 187,
    207, 213
  *Letters Written During a Short Residence
    in Sweden, Norway and Denmark*,
    55, 57, 59, 76–81, 191
  Mary, 56, 59, 81, 175, 190, 192–201,
    202
  Thoughts on the Education of Daughters,
    199

A Vindication of the Rights of Men, 183
A Vindication of the Rights of Woman,
    52–3, 56–9, 81, 122, 187, 190,
    199, 203
The Wrongs of Woman, 57, 176, 185,
    190, 191, 192, 194, 195, 198–206
Wood, Marcus, 171, 173

Wordsworth, William, 8, 128, 149, 154,
    156–7, 170, 214–15
Wright, Julia, 187

Young, Edward, 176, 194, 197

Žižka, John, 125, 132, 136